THE METAMORPHOSIS OF PLANTS

GOETHE.

THE METAMORPHOSIS OF PLANTS

Johann Wolfgang von Goethe

Introduction and photography by Gordon L. Miller

The MIT Press
Cambridge, Massachusetts
London, England

© 2009 Massachusetts Institute of Technology

All rights reserved. No part of this book may be reproduced in any form by any electronic or mechanical means (including photocopying, recording, or information storage and retrieval) without permission in writing from the publisher.

MIT Press books may be purchased at special quantity discounts for business or sales promotional use. For information, please e-mail special_sales@mitpress.mit.edu.

This book was set in Garamond Pro by The MIT Press.
Printed and bound in Spain.

Library of Congress Cataloging-in-Publication Data

Goethe, Johann Wolfgang von, 1749–1832.
The metamorphosis of plants / Johann Wolfgang von Goethe ; introduction and photography by Gordon L. Miller.
  p. cm.
 Includes bibliographical references and index.
 ISBN 978-0-262-01309-3 (alk. paper)
1. Plant morphology. 2. Geothe, Johann Wolfgang von, 1749–1832—Knowledge—Botany. I. Miller, Gordon L., 1954– II. Title.
  QK641.G599 2009
  571.8′ 762—dc22

                                                      2008044260

10 9 8 7 6 5 4 3 2

FRONTISPIECE (FIGURE A): Portrait of Goethe by C. A. Schwerdgeburth (1832)

I cannot tell you how readable the book of nature is becoming for me; my long efforts at deciphering, letter by letter, have helped me; now all of a sudden it is having its effect, and my quiet joy is inexpressible.

—Goethe to Charlotte von Stein, 1786

Contents

Preface
xi

Introduction
xv

The Metamorphosis of Plants
(poem)
1

The Metamorphosis of Plants
5

Appendix
The Genetic Method
105

Sources
117

Index
119

Preface

This little book has a rather ambitious goal—to promote not only greater but also deeper knowledge of the natural world. Johann Wolfgang von Goethe envisioned a fuller integration of poetic and scientific sensibilities that would provide a way of experiencing nature both symbolically and scientifically, simultaneously. *The Metamorphosis of Plants* represents Goethe's attempt to advance the scientific understanding of plants through such an integration at around the turn of the nineteenth century. For me personally, this edition has grown not only out of a lifelong fascination with plants but also from my interest in the relationship between Romanticism and modern science, and from my belief that Goethe's way of science offers hope for lessening the modern alienation from nature that not only diminishes the beauty and joy of human life but also fuels environmental irresponsibility and apathy.

A more specific stimulus for undertaking this newly illustrated edition of *The Metamorphosis of Plants* came from my teaching experience in the History Department and Environmental Studies Program at Seattle University. Having often introduced basic Goethean scientific ideas in my classes, I have been pleased to discover over the years, as Goethe also reported in 1817, that this "method of interpretation had captured young minds." I have found as well that graphic images are an indispensable aid in this educational effort and decided that I needed a more complete set. I was also eager to pursue a project that presented an opportunity to combine my intellectual interests with my photographic skills.

The most challenging part of the project, of course, was locating suitable specimens to photograph. Goethe mentions in his text some fifty different plants by genus or species, and I was able to find the majority of these. For those I could not find, I have drawn on some of the previously, and partially, illustrated editions of his book, as indicated in the list of sources. Goethe also often discusses a particular aspect of plant development without identifying a specific plant, so in many of these cases I have supplied an illustrative example.

The search for specimens led me far and wide around Western Washington State and beyond. I located some in the wild, I cultivated several myself, and I found many at various nurseries in the Puget Sound area. A few plants, however, were not to be found. Water buttercup (*Ranunculus aquatilis*), supposedly a fairly common plant in my area, proved surprisingly elusive. No nurseries currently carried it, and none of the local lakes or waterways seemed to harbor any. I finally found some after an hour-long chaperoned canoe trip around a lake in the highly protected Cedar River Watershed, the source of Seattle's drinking water, but it was the wrong variety. I ended up using an earlier line drawing for this illustration (see figure 3).

Another challenging specimen was a proliferous rose, which refers to a condition rather than a species. This phenomenon, which Goethe describes in paragraphs 103 and 104, is well-known to rosarians, but it occurs only under certain climatic or environmental conditions (which accords well with his theory of metamorphosis). I looked at lots and lots of roses—in nurseries, on street corners, and in the yards of neighbors and strangers, as well as in rose gardens in Seattle, Tacoma, Portland, San Francisco, and Minneapolis. At the Antique Rose Farm northeast of Seattle there was an excellent one, which bloomed, unfortunately, about a month before I arrived. This failure to find such a rose turned into an opportunity to include one of the several watercolors that were prepared for an earlier edition of Goethe's essay (see figure 19).

There have been a number of English translations of *Die Metamorphose der Pflanzen* since its original publication over two hundred years ago. The translation chosen for the present work is generally recognized as the modern standard, by Douglas Miller, which is contained in Volume 12 of *Goethe's Collected Works* published by Suhrkamp and is reprinted here with the permission of the publisher.

I wish to express my heartfelt thanks to the numerous helpful and knowledgable staff members at the many nurseries I frequented for their aid in finding particular plants or their advice on where I might. In addition, Dr. Arthur Kruckeberg eagerly offered helpful bits of his vast botanical knowledge as well as lively and encouraging discussions of Goethe. Arthur Lee Jacobson, both personally and through his excellent books on Seattle area plants, showed me the way to a number of species. Peg Pearson of the Washington Native Plant Society generously assisted my search for water buttercup and led me to Clay Antieau, a botanist with Seattle Public Utilities, who paddled with me around Walsh Lake on a glorious August day. Kevin Dann provided kind encouragement and welcome insights on both the text and the larger context of Goethean science. Friends and colleagues from Seattle University graciously offered invaluable help. Cordula Brown eagerly and skillfully helped me with some German sources while Dan Dombrowski, Erik Olsen, Trileigh Tucker, and Josef Venker provided excellent suggestions on the Introduction.

Preface

My editors at the MIT Press, particularly Bob Prior, Susan Buckley, and Sandra Minkkinen, have been most helpful, professional, and understanding throughout the metamorphosis of the book, and Yasuyo Iguchi created a design as beautiful as the plants themselves. My sister-in-law Debi Whisnant supplied a missing specimen by shipping a sprouting potato from Tennessee. And my wife, Jacquelyn, joined the search for several specimens (most memorably the high-altitude coltsfoot), improved my writing and offered insights on many images, and, above all, encouraged and endured my extended work on this project with heartening faith and abiding love. I am grateful to you all, and I readily take responsibility for any remaining errors that even all your good help could not keep me from making.

Introduction

Johann Wolfgang von Goethe remarked in his later years that he experienced the happiest moments of his extraordinary life during his devoted study of the metamorphosis of plants.[1] These gratifying and defining moments occurred largely during his sojourn in Italy from 1786 to 1788, a time when he was already famous as a writer both in his native Germany and abroad but was turning his prodigious abilities ever more intently to the scientific study of the natural world. This botanical research amid the lush Italian vegetation, as well as at home in the harsher German climate, resulted in a modest book first published in 1790 with the rather cautious title of *Attempt to Explain the Metamorphosis of Plants*. This work, whose size belies its significance, marked a turning point in Goethe's own intellectual life, and, in the words of historian Robert J. Richards, "seeded a revolution in thought that would transform biological science during the nineteenth century."[2] The text of this seminal scientific study, rendered into English, is the text before you now.

Because of Goethe's great renown as the author of the monumental *Faust* and other literary classics, it no doubt comes as a surprise to many to learn that he considered his scientific research and writing, diligently pursued through five decades, to be his most significant achievement. Beyond his work in botany, Goethe's scientific pursuits ranged from geology and meteorology to zoology and especially physics, wherein his sustained study of physiological optics resulted in his most substantial scientific treatise, his 1810 *Theory of Color*. Also in the opening decades of the nineteenth century, as he was drawing together various strands of his scientific investigations, Goethe coined the term and founded the fertile field of "morphology," a science of organic forms and formative forces aimed at discovering underlying unity in the vast diversity of plants and animals. He was also an insightful student of the history and philosophy of science and wrote many short essays on what he saw as the pitfalls and promise of modern scientific practice.

During much of his early life, Goethe gave little thought to the ways of nature, although from his youth he did have a sense of reverence for the irreducible and perhaps divine life of the natural world. But having grown up and been educated in large European cities, he was intellectually oriented toward the fashions of human society and the entertainments of polite literature, and he produced poetry that limned the inner passions of the human heart. It was not until 1775, when at the age of twenty-six he accepted a position in the court of Duke Charles Augustus at Weimar, that he exchanged "the stuffiness of town and study for the pure atmosphere of country, forest, and garden."[3] In this fresh new environment, as he carried out his administrative duties of overseeing the mines, the roads, the parks, the forests, and many other aspects of the duchy, Goethe began a disciplined inquiry into the natural order. His particular interest in plants burgeoned in the spring of 1776 when he began the regular planting and husbandry of a garden given to him by the duke. He schooled himself in the botanical classics, especially those of Linnaeus, whom he studied devotedly and sometimes daily. A decade spent in the stimulating air of the Weimar flora thus fortified Goethe's knowledge of plant life, but it was during his journey through Italy that he was seized by the crucial ideas that would inform his botanical investigations for the rest of his life.

While in Italy Goethe became convinced that he could discover some simple unity among the great variety of vegetation, an original or archetypal plant—an *Urpflanze*. There must be such an entity, he believed, otherwise, "how could I recognize that this or that form *was* a plant if all were not built upon the same basic model?"[4] At first he thought it might be possible to actually find this primal plant growing in some Mediterranean meadow or clinging to a rocky hillside. He gradually came to realize, however, that locating the *Urpflanze* would require looking in a much different place and in a qualitatively different way. Goethe had been fascinated with the progressive structure of the leaves of various plants, first of a palm tree in the Botanical Garden at Padua (samples of which he then carried around Italy and treasured for the rest of his life), and later of a fennel plant in Sicily, both of which suggested to him a unity of form in diverse structures (see image 6 and image 57). But he gained an insight central to his concept of metamorphosis while walking in the Sicilian gardens at Palermo. He says that "it came to me in a flash that in the organ of the plant which we are accustomed to call the *leaf* lies the true Proteus who can hide or reveal himself in all vegetal forms. From first to last, the plant is nothing but *leaf*, which is so inseparable from the future germ that one cannot think of one without the other."[5] The process through which this dynamic "leaf" progressively assumes the form of cotyledons, stem leaves, sepals, petals, pistil, stamens, and so on, is what Goethe meant by "the metamorphosis of plants."

What Goethe was discerning with this insight in Palermo was a deeper dimension in plant life, the realm of the "supersensuous plant archetype" lying beyond the empirically visible, touchable, smell-able, classifiable plant, undergirding and guiding the formation and transformation of the material shapes we see on the stem.[6] And his extensive empirical forays helped to persuade him that recognition of this dimension was necessary to account both for the apparent oneness in the great multitude of different plants and for the similarity of structure in the different parts of a single plant. In pursuing this approach to understanding the floral realm, Goethe was informed not only by his own insights and distinctive empiricism (discussed below) but also by ideas of the seventeenth-century Dutch philosopher Baruch Spinoza.

Goethe echoed Spinoza's holistic vision of reality in his conviction that "spirit and matter, soul and body, thought and extension . . . are the necessary twin ingredients of the universe, and will forever be."[7] And in order for us to comprehend not only the outer material aspect but also the inner, ideal, or archetypal aspect of natural things, Goethe discovered that we correspondingly must employ both the eyes of the body and the "eyes of the mind," both sensory and intuitive perception, "in constant and spirited harmony".[8] Goethe was especially struck by Spinoza's proposition that "the more we understand particular things, the more we understand God," and he coupled rigorous empiricism with precise imagination to see particular natural phenomena as concrete symbols of the universal principles, organizing ideas, or inner laws of nature. Starting from sense perception of the outer particulars, Goethe's scientific approach seeks the higher goal of an illuminating knowledge from within. This way of knowing—from the inside—is rooted ultimately in a harmony or identity between the human spirit and the informing spirit of nature, wherein "speaks one spirit to the other" (*Faust,* line 425).

There were good empirical reasons for describing the fundamental organ of the plant as a "leaf," and other botanists before Goethe

Introduction

(unbeknownst to him at the time) had proposed similar theories. For one thing, stem or foliage leaves often offer ready evidence of transitions in plant development, effectively anticipating in their structure or coloration a subsequent stage, and thus were seen as abiding close to the fundamental formative process. He also emphasized, however (see paragraphs 119–121), that one could view these transitions from starting points other than stem leaves and could profitably envision the metamorphic process going backward as well as forward. Thus we could, for example, see a sepal as a contracted stem leaf, or a stem leaf as an expanded sepal; a stamen as a contracted petal, or a petal as an expanded stamen. While *leaf* is a common, convenient, and opportune term, the crucial concept coursing and pulsing throughout Goethe's botany is that of the dynamic inward archetype, which we can conceive as a vibrant field of formative forces and which he dubbed the "true Proteus." This central theme of the protean character of the ideal organ of the leaf informs the whole of *The Metamorphosis of Plants*. It is therefore important to remember, while reading the particulars of the text, that Goethe's overall intent was for the parts to form a whole and fluid story of floral forms in process—to present, in effect, a motion picture of the metamorphosis of plants.

The notion of metamorphosis had long been applied to the transformation of caterpillars into butterflies and tadpoles into frogs, which Goethe had also carefully studied. By extending this concept to the development of plants, he was suggesting the presence of a lawful process working in various ways throughout the realms of nature. He later specified two aspects to this ordered but productive power—"two great driving forces in all nature"—which he identified as "intensification" and "polarity." *Intensification* is "a state of ever-striving ascent" toward greater complexity or perfection, toward the fullest possible expression in physical, empirical phenomena of the potential inherent in the underlying idea or *Urphenomen*. Goethe saw evidence of intensification in the metamorphosis of a plant from cruder, simpler, and vegetative stem leaves to finer and more colorful petals and specialized reproductive organs. The fulfillment of the process, he argued, requires the progressive refinement of sap by successive plant structures.

The kindred concept of *polarity* is, in its most basic form (such as in the domain of electricity and magnetism), "a state of constant attraction and repulsion" that more generally involves a dynamic and creative interplay of opposites. In the metamorphosis of plants, polarity is most evident in the alternating forces of expansion and contraction Goethe identified in the stages of development. In paragraph 73 he outlines six stages in this polar process—expansion from the seed into stem leaves, contraction from stem leaves into the sepals of the calyx, expansion from sepals into petals, contraction from petals into pistil and stamens, expansion from reproductive organs into fruit, and, completing the cycle, contraction from fruit into seed. Through these steps, "nature steadfastly does its eternal work of propagating vegetation by two genders."

The Faustian striving of natural things in the process of intensification, in the alternating rhythm of nature's grand systole and diastole, represented for Goethe a universal impulse ascending "as on a spiritual ladder" (paragraph 6) through relatively unformed matter to more complete manifestations of the nonmaterial ideas at the heart of things. This ladder, however, was not so rigid as to always end at a predetermined point. The characteristic expression of an underlying idea in any particular plant was, for Goethe, always the coordinated result of "the law of inner nature, whereby the plant has been constituted" and "the law of environment, whereby the plant has been modified." The development of organic forms always proceeds, therefore, both "from within toward without" and "from without toward within."[9] He mentions specifically, for example, how leaf structure can be affected by the relative wetness or dryness of habitats, often at different elevations (paragraphs 24–25), and how excessive nourishment can retard flowering (paragraphs 30, 38, and 109).

As an aid to understanding Goethe's overall approach to metamorphosis throughout nature, we can distinguish three closely related aspects of the process, which I have expressed here in terms of his central notion of the Proteus. There is, first of all, the basic nucleus of formative forces with its rich productive potential—*Proteus in potentia.* Then there is the actualization of this inner potential in a diverse range of organic forms—leaves, petals, pistils, backbones, and blue-footed boobies—*Proteus actus.* These actual physical structures and qualities, however, are affected by changing external conditions, which means that the preceding notion must always carry a qualifier—*Proteus actus adaptatus*—the formative potential actualized but adapted to its environment.

The process of dialectical development envisioned by Goethe helps fashion the mutual "fitness" of organism and environment, but he did not see this process as fulfilling any predesigned purposes or aiming for any fixed ends. Rejecting the classical notion of external teleology in nature, he proposed that we can "attain a more satisfactory insight into the mysterious architecture of the formative process" if we study "how nature expresses itself from all quarters and in all directions as it goes about its work of creation."[10] For Goethe, the integrity and rising intensity of the inner impulse, the creativity of which sometimes issues in complexities of form far beyond the needs of mere survival, gives natural things a degree of autonomy and a measure of intrinsic value. They, and nature in toto, are destined not for particular—and particularly anthropocentric—ends, but rather are striving for the internal satisfaction of wholeness. Indeed, Goethe's emphasis on the interdependence of organism and environment, as well as organism and organism—"in which one species is created, or at least sustained, by and through another"—presents a view that can surely be described as ecological, seventy-five years before German biologist Ernst Haeckel coined the term.[11] Haeckel in fact, who was a tireless promoter of Darwin's theory of evolution, was also an assiduous champion of Goethe as an evolutionary precursor. Darwinian natural history, however, is metaphysically more limited than Goethe's. In relation to the terminology suggested above, it aims to interpret *actus adaptatus* absent the ideal Proteus with its creative potential and inner law.

In the latter months of his Italian journey, Goethe wrote his philosopher friend J. G. Herder saying, "I believe I have come very close to the truth about the *how* of the organism."[12] Upon returning to Germany in the spring of 1788, he continued to ponder and discuss these insights and ideas with his Weimar circle. Then, in the following year and a half, he succeeded in systematizing his thinking and, in the manner of Linnaeus's great works, setting it down in a series of 123 numbered paragraphs. The book appeared at Easter of 1790, the author's first scientific publication.

Even though Goethe had steeped himself for years in the works of Linnaeus and had immense respect, and even reverence, for the great taxonomist, his own botanical work represented a departure from the Swede's scientific approach. Goethe recognized the value of systematic classification in bringing a sense of order to the teeming multitudes of flora and fauna, but he also felt constrained by the limits of Linnaeus. Aside from finding the practice of naming and enumerating the parts of plants and summing them into a whole to be rather artificial and mechanical, his primary difficulty with the Linnaean system was that he found the terminology inadequate to accommodate the variability of organs, whether the differential leaf structure serially displayed on single stems or the different forms of plants of the same species growing under different conditions. Considering this variability, Goethe decided that it would be fruitless to search among these multifarious forms for the enduring essence of plant life. It must reside instead in the realm of dynamic archetypes, his recognition of which propelled him beyond the Linnaean world of fixed forms and species and into a new world of transformation and evolution.

Because supersensible archetypes or objective ideas in nature are not things recognized by mainstream modern science, many have been led to reject Goethe's scientific approach as suffering too much from the romantic musings of his poetic genius. But to aim this criticism at Goethe's way of science is merely to beg the question he was posing about the limits of mainstream science: Can a mechanistic, materialistic approach that focuses only on innumerable individual surface structures meet the explanatory challenge of the living organism or the life of nature as a whole? His sense that the world we experience could never be built up from mere matter in motion, nor truly known on the model of a human subject confronting a mere object, spurred him to develop his alternative approach. In contrast to conventional empiricism, Goethe advocated a "delicate empiricism which makes itself utterly identical with the object."[13] This mode of inquiry aims to overcome subject/object dualism by endowing detailed sense experience of the outward forms of nature with the enlivening inward power of imagination, while also grounding subjective imagination in objective forms and facts. So, in place of the alienation from the natural world at the center of the conventional Cartesian approach, Goethe proposed a way of identification as the path to a deeper and unifying knowledge of nature.

Goethe embodied his belief that science and poetry, with their corresponding conceptions of nature, are not incompatible but actually complementary. The poet-scientist, however, has often struggled for a hearing in the modern Western world. Goethe was bewildered by many of the responses his essay engendered among his circle of friends and acquaintances. What was this new effort supposed to be?

It seemed too scientific for poetry, but perhaps also too poetic for science. "Nowhere," he complained, "would anyone grant that science and poetry can be united. People forgot that science had developed from poetry and they failed to take into consideration that a swing of the pendulum might beneficently reunite the two, at a higher level and to mutual advantage."[14] Years later Goethe composed a poem also titled "The Metamorphosis of Plants" (which follows this Introduction) in an effort to make his scientific theories and pursuits more palatable and intelligible to his wife and women friends, though with only limited success. His scientific treatise did, however, receive three very favorable reviews in German periodicals soon after its appearance, as well as supportive references in a variety of botanical publications.

More weighty and considered appreciation of Goethe's work in botany appeared throughout the first half of the nineteenth century. The great naturalist Alexander von Humboldt dedicated an 1806 book to Goethe with an illustration featuring *The Metamorphosis of Plants* and imagery suggesting, true to Humboldt's Romantic sympathies, that poetry as well as science can succeed in uncovering the secrets of nature. In 1853 renowned physicist and physiologist Hermann von Helmholtz praised Goethe's theory of metamorphosis in plants and animals. He reiterated his praise forty years later, in 1892, and at that point, after the publication of *On the Origin of Species* in 1859, said that Goethean morphology had so shaped nineteenth-century biology that it paved the way for Darwin's theory. Darwin himself, in fact, made several direct references to Goethe's theories of metamorphosis in various works, including the *Origin*. Robert J. Richards has recently reviewed and reinforced these lines of influence by proposing that "evolutionary theory was Goethean morphology running on geological time."[15] In this evolution of ideas, however, as indicated above, the Proteus would turn prosaic, as Goethe's ideal archetypes would be reduced to material, historical, ancestral creatures.

Other leading nineteenth-century thinkers more readily accepted both the material and the ideal sides of Goethe's science. French naturalist Étienne Geoffroy Saint-Hilaire and Swiss botanist Augustin Pyrame de Candolle each developed forms of idealistic, or "transcendental," biology during the early decades of the century. And in England, toward mid-century, eminent zoologist Richard Owen developed archetypal ideas into a major theory of evolution, which he defended vigorously against the Darwinian view of evolution by merely material forces.

Across the Atlantic, the American Transcendentalists Ralph Waldo Emerson and Henry David Thoreau embraced Goethe enthusiastically and valued especially *The Metamorphosis of Plants*. Goethe's botany contributed directly to Emerson's sense of the unity and progressive dynamism of nature, to his ideal image of science as a search for the "pure plastic idea," and thus to his vision of a morality grounded in nature and rising through universal energies to the heights of human potential. Thoreau's debt to Goethe is evident in much of his work but is perhaps most explicit in the *Walden* chapter entitled "Spring." Here, in his justly famous observations on the flowing forms of sand in a thawing bankside, his transcendental vision is sparked by an inkling of vegetation:

> You find thus in the very sands an anticipation of the vegetable leaf. No wonder that the earth expresses itself outwardly in leaves, it so labors with the idea inwardly. The atoms have already learned this law, and are pregnant by it. The overhanging leaf sees here its prototype. . . . The feathers and wings of birds are still drier and thinner leaves. . . . Even ice begins with delicate crystal leaves . . . [and] the whole tree itself is but one leaf . . . Thus it seemed that this one hill-side illustrated the principle of all the operations of Nature. The Maker of this earth but patented a leaf. What Champollion will decipher this hieroglyphic for us, that we may turn over a new leaf at last?"[16]

The *Urpflanze* had indeed become well rooted in a particularly hospitable patch of American soil.

Considering the mixed response to *The Metamorphosis of Plants* in his lifetime, Goethe eventually came to realize that his scientific aspirations lay "entirely outside the intellectual horizon of the time."[17] The intellectual horizon of *our* time apparently has, by and large, many of the same coordinates—the pendulum has not yet swung sufficiently to reunite science and poetry, and Goethe's delicate empiricism is still, if anything, "alternative."[18] *The Metamorphosis of Plants*, however, is first among Goethe's scientific works in terms of favorable, though not unqualified, reception from the modern scientific community. His basic proposition that "all is leaf," commonly known as the foliar theory, has in fact, in the words of a recent text, "underpinned all work on flower development, including modern molecular genetic analysis."[19] The genetic work of distinguished biologists Enrico Coen, Elliot Meyerowitz, and others is particularly significant for providing experimental support for this guiding insight of Goethe's, as well as for his view that floral abnormalities can reveal the inner workings of normal development.[20] Other details of Goethe's science, as with any scientific theory, have been variously accepted or assailed. The essential significance of his scientific endeavors, however, lies not in the sum of factual knowledge he contributed but rather in the way of knowing he developed. Yet his method, with its associated metaphysics, has been generally less accepted than his facts.

There are nevertheless some encouraging signs, especially in recent decades, as a growing number of scientists from a range of fields have looked to Goethe for ideas and inspiration. To take just one example, Swiss biologist Adolf Portmann, believing that "Goethe's image of the metamorphosis of plants has placed before our eyes the grandeur of living nature," has followed Goethe's lead in various directions and has promoted in this spirit a new kind of science that "leads to a deepened

Introduction

experience with the realm of living forms and makes nature for us a true home." Aside from Goethe's potential value for contemporary science and scientists, a rigorously empirical approach to nature that can also bring us more deeply home to the biosphere certainly seems, in the face of current environmental concerns, like an attractive alternative.[21]

Goethe described *The Metamorphosis of Plants* as a "short preliminary treatise" (paragraph 9), which he fully intended to expand into a more substantial and convincing sequel. He made a start on this sequel, but it never came to full fruition. Because Goethe placed great stock in the significance of visual images for both the advancement of science and the conduct of life, one of the most important additions he planned for the sequel was the inclusion of illustrations. He had a number of these prepared, including the examples below.

 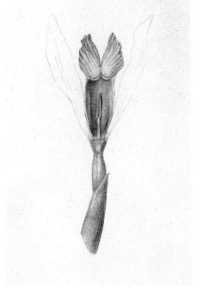

FIGURE B AND FIGURE C: Watercolors commissioned by Goethe in the early 1790s for a sequel to *The Metamorphosis of Plants*

This immortal genius of world literature even once went so far as to say that "[w]e ought to talk less and draw more. I, personally, should like to renounce speech altogether and, like organic nature, communicate everything I have to say in sketches."[22] Graphic images also figure prominently in the way of knowing that Goethe called "exact sensory imagination," by which one might penetrate the surface of things and gain the depths.[23] Many of the original illustrations for the essay, a large number of which are limited to only the first stage of plant development, were eventually published, and several partially and variously illustrated editions have appeared over the past two hundred years. The present work, however, is the most thoroughly and colorfully illustrated edition, and the only one to be illustrated photographically.[24]

Goethe thought of illustrations as often beneficially standing in lieu of nature and was thus well aware of the necessity of faithful depiction of natural objects. So he recognized the importance in all natural history illustration for artists to abide by the canons of light and shadow and the rules of perspective, or perhaps to utilize the newly invented camera lucida or camera clara, to ensure an accurate representation. He thus wished to join, in his life and work, not only *poetry* and science, but *art* and science as well. Early in the nineteenth century, he was sanguine that his botanical book would one day be effectively illustrated as he saw great opportunities for the advancing science to be well served by improved graphic techniques. I trust he would be pleased with the results made possible by the further evolution of the camera in the service of his idealistic scientific vision.

Goethe always wanted his wide-ranging scientific work to reach an audience beyond the domain of botanists, zoologists, and physicists. Almost thirty years after the publication of *The Metamorphosis of Plants,* he expressed disappointment at being unable himself to achieve his dream of a superior sequel. His aim, as he described it, was "to do nothing less than to present to the physical eye, step by step, a detailed, graphic, orderly version of what I had previously presented to the inner eye conceptually and in words alone, and to demonstrate to the exterior senses that the seed of this concept might easily and happily develop into a botanical tree of knowledge whose branches might shade the entire world."[25] It is my hope that the present illustrated edition, while not the full sequel that Goethe envisioned, will nevertheless aid the metamorphosis of that tree of knowledge toward diverse and wide-spreading foliage and particularly deep roots.

NOTES

1. See Goethe, *Scientific Studies,* edited and translated by Douglas Miller (New York: Suhrkamp, 1988), 18.
2. Robert J. Richards, *The Romantic Conception of Life: Science and Philosophy in the Age of Goethe* (Chicago: University of Chicago Press, 2002), 407.
3. Goethe, "The Author Relates the History of His Botanical Studies," in *Goethe's Botanical Writings,* trans. by Bertha Mueller. (Honolulu: University Press of Hawaii, 1952; reprint, Ox Bow Press, 1989), 150.
4. Goethe, *Italian Journey* (London: Penguin, 1962), 258–259.
5. Ibid., 366. Proteus was a Greek mythological god of the sea who could assume different shapes at will.
6. Goethe, "The Author Relates the History of His Botanical Studies," in *Goethe's Botanical Writings,* 162.
7. Goethe to Karl Ludwig von Knebel (8 Aug. 1812), quoted in Ludwig Lewisohn, *Goethe: The Story of a Man,* Vol. 2 (New York: Farrar, Straus and Co., 1949), 200.
8. Goethe, "My Discovery of a Worthy Forerunner," in *Goethe's Botanical Writings,* 180. See the Appendix for further elaboration of this approach.
9. Goethe, "Preliminary Notes for a Physiology of Plants," in *Goethe's Botanical Writings,* 83, 85.
10. Goethe, "Toward a General Comparative Theory," in *Scientific Studies,* 55, 56.
11. Ibid., 56
12. Goethe, *Italian Journey,* 379.

Introduction

13. Goethe, *Maxims and Reflections,* in *Scientific Studies,* 307.
14. Goethe, "History of the Printed Brochure," in *Goethe's Botanical Writings,* 171–172.
15. Richards, *Romantic Conception of Life,* 407.
16. Henry D. Thoreau, *Walden,* ed. by Jeffrey S. Cramer (New Haven: Yale University Press, 2004), 295–297.
17. Goethe, "Other Friendly Overtures," in *Goethe's Botanical Writings,* 185.
18. For authoritative discussions of Goethe's science in relation to mainstream science, see *Goethe and the Sciences: A Reappraisal,* ed. by Frederick Armine, Francis J. Zucker, and Harvey Wheeler (Dordrecht: D. Reidel, 1987) and *Goethe's Way of Science,* ed. by David Seamon and Arthur Zajonc (Albany: State University of New York Press, 1998). See also Rudolf Steiner's insightful *Goethe's World View* (Spring Valley, N.Y.: Mercury, 1985).
19. Beverley J. Glover, *Understanding Flowers and Flowering* (Oxford: Oxford University Press, 2007), 10.
20. Enrico Coen, *The Art of Genes: How Organisms Make Themselves* (Oxford: Oxford University Press, 1999), esp. Chap. 4. Stephen J. Gould, in *The Structure of Evolutionary Theory* (Cambridge: Harvard University Press, 2002), presents an appreciative critique of Goethe's botanical work (pp. 281–91) and summarizes many of the recent experimental validations of the basic foliar theory (pp.1092–95). See also Gould's "More Light on Leaves" (*Natural History* 100, Feb. 1991, 16–23). Goethe specifically notes the value of abnormalities in paragraphs 3, 7 and 41.
21. Adolf Portmann, "Goethe and the Concept of Metamorphosis," in *Goethe and the Sciences,* 144, 145. Portmann's student Andreas Suchantke extends the Goethean scientific approach to the perception of landscape in *Eco-Geography* (Great Barrington, Mass.: Lindisfarne, 2001) and the educational programs of The Nature Institute (natureinstitute.org) promote "delicate empiricism" in a variety of contexts.
22. Goethe, *Italian Journey,* 9.
23. See the Appendix for a description of "exact sensory imagination."
24. Various presentations of the original watercolor illustrations are contained in the following works: *Goethes Metamorphose der Pflanzen,* edited by Adolph Hansen (Giessen: Töpelmann, 1907); Goethe, *Die Metamorphose der Pflanzen: Mit dem Originalbildwerk,* edited by Julius Schuster (Berlin: W. Junk, 1924); Goethe, *Die Schriften zur Naturwissenschaft,* Vol. 9 (*Morphologische Hefte*), edited by Dorothea Kuhn (Weimar: Böhlaus, 1954); and Goethe, *Die Metamorphose der Pflanzen,* edited by Dorothea Kuhn (Weinheim: Acta Humaniora, 1984). Goethe's original line drawings

can be found in *Corpus der Goethezeichnungen,* Vol. 5B, edited by Dorothea Kuhn, Otfried Wagenbreth, and Karl Schneider-Carius (Leipzig: Seemann, 1967). Supplemental drawings, many of which are included in the present edition, are integrated into the text of *Die Metamorphose der Pflanzen* contained in *Goethes Morphologische Schriften,* edited by Wilhelm Troll (Jena: Diederichs, 1926, 1932).

25. Goethe, "Later Studies and Collections," in *Goethe's Botanical Writings,* 97.

The Metamorphosis of Plants
(Goethe poem)

The rich profusion thee confounds, my love,
Of flowers, spread athwart the garden. Aye,
Name upon name assails thy ears, and each
More barbarous-sounding than the one before—
Like unto each the form, yet none alike;
And so the choir hints a secret law,
A sacred mystery. Ah, love could I vouchsafe
In sweet felicity a simple answer!
Gaze on them as they grow, see how the plant
Burgeons by stages into flower and fruit,
Bursts from the seed so soon as fertile earth
Sends it to life from her sweet bosom, and
Commends the unfolding of the delicate leaf
To the sacred goad of ever-moving light!
Asleep within the seed the power lies,
Foreshadowed pattern, folded in the shell,
Root, leaf, and germ, pale and half-formed.
The nub of tranquil life, kept safe and dry,

The Metamorphosis of Plants (Poem)

Swells upward, trusting to the gentle dew,
Soaring apace from out the enfolding night.
Artless the shape that first bursts into light—
The plant-child, like unto the human kind—
Sends forth its rising shoot that gathers limb
To limb, itself repeating, recreating,
In infinite variety; 'tis plain
To see, each leaf elaborates the last—
Serrated margins, scalloped fingers, spikes
That rested, webbed, within the nether organ—
At length attaining preordained fulfillment.
Oft the beholder marvels at the wealth
Of shape and structure shown in succulent surface—
The infinite freedom of the growing leaf.
Yet nature bids a halt; her mighty hands,
Gently directing even higher perfection,
Narrow the vessels, moderate the sap;
And soon the form exhibits subtle change.
The spreading fringes quietly withdraw,
Letting the leafless stalk rise up alone.
More delicate the stem that carries now
A wondrous growth. Enchanted is the eye.
In careful number or in wild profusion
Lesser leaf brethren circle here the core.
The crowded guardian chalice clasps the stem,
Soon to release the blazing topmost crown.
So nature glories in her highest growth,
Showing her endless forms in orderly array.
None but must marvel as the blossom stirs
Above the slender framework of its leaves.
Yet is this splendor but the heralding

The Metamorphosis of Plants (Poem)

Of new creation, as the many-hued petals
Now feel God's hand and swiftly shrink. Twin forms
Spring forth, most delicate, destined for union.
In intimacy they stand, the tender pairs,
Displayed about the consecrated altar,
While Hymen hovers above. A swooning scent
Pervades the air, its savor carrying life.
Deep in the bosom of the swelling fruit
A germ begins to burgeon here and there,
As nature welds her ring of ageless power,
Joining another cycle to the last,
Flinging the chain unto the end of time—
The whole reflected in each separate part.
Turn now thine eyes again, love, to the teeming
Profusion. See its bafflement dispelled.
Each plant thee heralds now the iron laws.
In rising voices hear the flowers declaim;
And, once deciphered, the eternal law
Opens to thee, no matter what the guise—
Slow caterpillar or quick butterfly,
Let man himself the ordained image alter!
Ah, think thou also how from sweet acquaintance
The power of friendship grew within our hearts,
To ripen at long last to fruitful love!
Think how our tender sentiments, unfolding,
Took now this form, now that, in swift succession!
Rejoice the light of day! Love sanctified,
Strives for the highest fruit—to look at life
In the same light, that lovers may together
In harmony seek out the higher world!

The Metamorphosis of Plants

Introduction

1

Anyone who has paid even a little attention to plant growth will readily see that certain external parts of the plant undergo frequent change and take on the shape of the adjacent parts—sometimes fully, sometimes more, and sometimes less.

2

Thus, for example, the single flower most often turns into a double one when petals develop instead of stamens and anthers; these petals are either identical in form and color to the other petals of the corolla, or still bear visible signs of their origin.

3

Hence we may observe that the plant is capable of taking this sort of backward step, reversing the order of growth. This makes us all the more aware of nature's regular course; we will familiarize ourselves with the laws of metamorphosis by which nature produces one part through another, creating a great variety of forms through the modification of a single organ.

4

Researchers have been generally aware for some time that there is a hidden relationship among various external parts of the plant that develop one after the other and, as it were, one out of the other (for example, leaves, calyx, corolla, and stamens); they have even investigated the details. The process by which one and the same organ appears in a variety of forms has been called *the metamorphosis of plants.*

5

This metamorphosis appears in three ways: *regular, irregular* and *accidental.*

6

Regular metamorphosis may also be called *progressive* metamorphosis: it can be seen to work step by step from the first seed leaves to the last formation of the fruit (image 1). By changing one form into another, it ascends—as on a spiritual ladder—to the pinnacle of nature: propagation through two genders. I have observed this carefully for several years, and now propose to explain it in the present essay. Hence, in the following discussion we will consider only the annual plant that progresses continuously from seed to fruiting.

7

Irregular metamorphosis might also be called *retrogressive* metamorphosis (image 2). In the previous case nature pressed forward to her great goal, but here it takes one or more steps backward. There, with irresistible force and tremendous effort, nature formed the flowers and equipped them for works of love;[1] here it seems to grow slack, irresolutely leaving its creation in an indeterminate, malleable state often pleasing to the eye but lacking in inner force and effect. Our observations of this metamorphosis will allow us to discover what is hidden in regular metamorphosis, to see clearly what we can only infer in regular metamorphosis. Thus we hope to attain our goal in the most certain way.

FIGURE 1: The annual plant, Goethe's basic model in his discussion of metamorphosis; plant parts, separated for the purpose of illustration, from top to bottom—pistil, stamens, corolla, calyx, stem leaves, cotyledons, and roots.

IMAGE 1: *Chrysanthemum morifolium* displaying regular metamorphosis

IMAGE 2: *Chrysanthemum grandiflorum* displaying irregular metamorphosis

8

We will, however, leave aside the third metamorphosis, caused *accidentally* and from without (especially by insects). It could divert us from the simple path we have to follow, and confuse our purpose. Opportunity may arise elsewhere to speak of these monstrous but rather limited excrescences.

9

I have ventured to develop the present essay without reference to illustrations, although they might seem necessary in some respects. I will reserve their publication until later; this is made easier by the fact that enough material remains for further elucidation and expansion of this short preliminary treatise. Then it will be unnecessary to proceed in the measured tread required by the present work. I will be able to refer to related matters, and several passages gleaned from like-minded writers will be included. In particular, I will be able to use comments from the contemporary masters who grace this noble science. It is to them that I present and dedicate these pages.

I. Of the Seed Leaves

10

Since we intend to observe the successive steps in plant growth, we will begin by directing our attention to the plant as it develops from the seed. At this stage we can easily and clearly recognize the parts belonging to it. Its coverings (which we will not examine for the moment) are left more or less behind in the earth, and in many cases the root establishes itself in the soil before the first organs of its upper growth (already hidden under the seed sheath) emerge to meet the light.

11

These first organs are known as *cotyledons;* they have also been called seed lobes, nuclei, seed laps, and seed leaves in an attempt to characterize the various forms in which we find them.

12

They often appear unformed, filled with a crude material, and as thick as they are broad. Their vessels are unrecognizable and scarcely distinguishable from the substance of the whole; they have little resemblance to a leaf, and we could be misled into considering them separate organs.

13

In many plants, however, they are more like the leaf in form. They become flatter; their coloration turns greener when they are exposed to light and air; and their vessels become more recognizable, more like the ribs of a leaf.

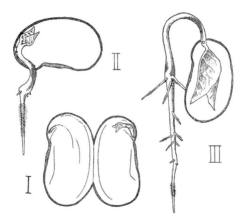

FIGURE 2: Germination of the garden bean: I Seed opened, cotyledons separated to reveal embryo; II Bean in process of germination, one cotyledon removed to reveal growing embryo, the latter now with strong root and radically increased in size; III Advanced stage of germination

The Metamorphosis of Plants

14

In the end they appear as real leaves: their vessels are capable of the finest development, and their resemblance to the later leaves prevents us from considering them separate organs. Instead, we recognize them as the first leaves of the stem.

15

But a leaf is unthinkable without a node, and a node is unthinkable without an eye. Hence we may infer that the point where the cotyledons are attached is the first true node of the plant. This is confirmed by those plants that produce new eyes directly under the wings of the cotyledons, and develop full branches from these first nodes (as, for example, in *Vicia faba*).

IMAGE 3: Broad bean seedling (*Vicia faba*)

The Metamorphosis of Plants

IMAGE 4: Broad bean seedling (*Vicia faba*)

16

The cotyledons are usually double, and here we must make an observation that will become more important later. The leaves of this first node are often paired, whereas the later leaves of the stem alternate; that is, here parts are associated and joined which nature later separates and scatters. Even more noteworthy is the appearance of the cotyledons as a collection of many small leaves around a single axis, and the gradual development of the stem from its center to produce the later leaves singly; this can be seen quite clearly in the growth of the various kinds of pines (image 5). Here a circle of needles forms something like a calyx—we will have occasion to remember this when we come to similar phenomena.

17

We will ignore for the moment the quite unformed, individual nuclei of those plants that sprout with but a single leaf.

18

We will, however, note that even the most leaflike cotyledons are always rather undeveloped in comparison to the later leaves of the stem. Their periphery is quite uniform, and we are as little able to detect traces of serration there as we are to find hairs on their surfaces, or other vessels[2] peculiar to more developed leaves.

II. Development of the Stem Leaves from Node to Node

19

Now that the progressive effects of nature are fully visible, we can see the successive development of the leaves clearly. Often one or more of the following leaves were already present in the seed, enclosed between the cotyledons; in their closed state they are known

Image 5: Austrian black pine (*Pinus nigra*)

as plumules. In different plants their form varies in relation to that of the cotyledons and the later leaves; most often they differ from the cotyledons simply in being flat, delicate, and generally formed as true leaves. They turn completely green, lie on a visible node, and are undeniably related to the following stem leaves, although they usually lag behind in the development of their periphery, their edge.

20

But further development spreads inexorably from node to node through the leaf: the central rib lengthens, and the side ribs along it reach more or less to the edges. These various relationships between the ribs are the principal cause of the manifold leaf forms. The leaves now appear serrated, deeply notched, or composed of many small leaves (in which case they take the shape of small, perfect branches). The date palm presents a striking example of such successive and pronounced differentiation in the most simple leaf form (image 6). In a sequence of several leaves, the central rib advances, the simple fanlike leaf is torn apart, divided, and a highly complex leaf is developed that rivals a branch.

21

The development of the leaf stalk keeps pace with that of the leaf itself, whether the leaf stalk is closely attached to the leaf or forms a separate, small, easily severed stalk.

22

In various plants we can see that this independent leaf stalk has a tendency to take on the form of a leaf (for example, in the orange family) (image 7). Its structure will give rise to certain later observations, but for the moment we will pass them by.

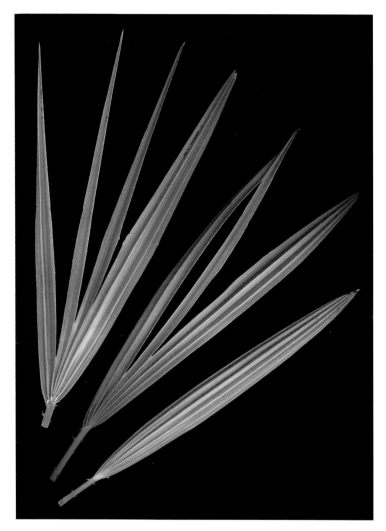

IMAGE 6: Leaves of Mediterranean fan palm (*Chamaerops humilis*) showing successive differentiation in form from bottom to top. This is the species of palm that drew Goethe's attention at the botanical garden in Padua, Italy.

IMAGE 7: Leaf of lime (*Citrus aurantiifolia*) with broadened petiole

23

Neither can we enter here into further consideration of the stipules; we will simply note in passing that they share in the later transformation of the stalk, particularly when they form a part of it.

24

Although the leaves owe their initial nourishment mainly to the more or less modified watery parts that they draw from the stem, they are indebted to the light and air for the major part of their development and refinement. We found almost no structure and form, or only a coarse one, in those cotyledons produced within the closed seed covering and bloated, as it were, with a crude sap. The leaves of underwater plants likewise show a coarser structure than those of plants exposed to the open air; in fact, a plant growing in low-lying, damp spots will even develop smoother and less refined leaves than it will when transplanted to higher areas, where it will produce rough, hairy, more finely detailed leaves (images 8A, 8B).

25

In the same way, more rarefied gases are very conducive to, if not entirely responsible for, the anastomosis[3] of the vessels that start from the ribs, find one another with their ends, and form the leaf skin. The leaves of many underwater plants are threadlike, or assume the shape of antlers; we are inclined to ascribe this to an incomplete anastomosis. This is shown at a glance by the growth of *Ranunculus aquaticus,* where the leaves produced underwater consist of threadlike ribs, although those developed above water are fully anastomosed and form a connected surface. In fact, we can see the transition clearly in the half-anastomosed, half-threadlike leaves found in this plant (figures 3, 4).

The Metamorphosis of Plants

IMAGES 8A AND 8B: Two varieties of coltsfoot displaying intraspecies differences in leaf morphology at different elevations. *Top: Petasites frigidus* var. *nivalis* from 5,000-foot elevation in the Cascade Mountains; *bottom: Petasites frigidus* var. *palmatus* from 2,400-foot elevation in the Cascade foothills

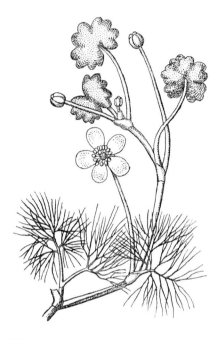

Figure 3: Water buttercup (*Ranunculus aquaticus*) displaying fully formed aerial leaves and threadlike submerged ones

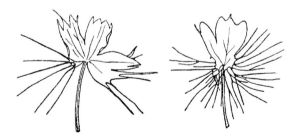

Figure 4: Floating leaves of water buttercup representing transitions to submerged leaves

26

Experiments have shown that the leaves absorb different gases, and combine them with the liquids they contain; there is little doubt that they also return these refined juices to the stem, and thereby help greatly in the development of the nearby eyes. We have found convincing evidence for this in our analysis of gases developed from the leaves of several plants, and even from the hollow stems.

27

In many plants we find that one node arises from another. This is easy to see in stems closed from node to node (like the cereals, grasses, and reeds), but not so easy to see in other plants that are hollow throughout and filled with a pith or rather, a cellular tissue. This substance, previously called *pith,* was considered to occupy an important position among the inner parts of the plant, but its importance has recently been disputed, and with good cause in my opinion (Hedwig, *Leipzig Magazine,* no. 3).[4] Its supposed influence on growth has been flatly denied; the force for growth and reproduction is now ascribed wholly to the inner side of the second bark, the so-called liber. Since the upper node arises from the node below and receives sap from it, we can easily see that the node above must receive a sap which is finer and more filtered; it must benefit from the effect of the earlier leaves, take on a finer form, and offer its own leaves and eyes even finer juices.[5]

28

As the coarser liquids are continually drawn off and the purer ones introduced, as the plant refines its form step by step, it reaches the point ordained by nature. We finally see the leaves in their maximum size and form, and soon note a new phenomenon that tells us that the previous stage is over and the next is at hand, the stage of the flower.

III. Transition to Flowering

29

The transition to flowering may occur quickly or slowly. In the latter case we usually find that the stem leaves begin to grow smaller again, and lose their various external divisions, although they expand somewhat at the base where they join the stem. At the same time we see that the area from node to node on the stem grows more delicate and slender in form; it may even become noticeably longer.

30

It has been found that frequent nourishment hampers the flowering of a plant, whereas scant nourishment accelerates it. This is an even clearer indication of the effect of the stem leaves discussed above. As long as it remains necessary to draw off coarser juices, the potential organs of the plant must continue to develop as instruments for this need. With excessive nourishment this process must be repeated over and over; flowering is rendered impossible, as it were. When the plant is deprived of nourishment, nature can affect it more quickly and easily: the organs of the nodes[6] are refined, the uncontaminated juices work with greater purity and strength, the transformation of the parts becomes possible, and the process takes place unhindered.

IV. Formation of the Calyx

31

We often find this transformation occurring rapidly. In this case the stem, suddenly lengthened and refined, shoots up from the node of the last fully formed leaf and collects several leaves around the axis at its end.

FIGURE 5: Original sketches by Goethe, showing development of stems from nodes and leaves. *Left:* contraction of stem leaves to the calyx; *center:* succession of nodes; *right:* node with leaf

32

The leaves of the calyx are the same organs that appeared previously as the leaves of the stem; now, however, they are collected around a common center, and often have a very different form. This can be demonstrated in the clearest possible way.

33

We already noted a similar effect of nature in our discussion of the cotyledon, where we found several leaves, and apparently several nodes, gathered together around one point. As the various species of pine develop from the seed, they display a rayed circle of unmistakable needles that, unlike other cotyledons, are already well developed. Thus in the earliest infancy of this plant we can already see a hint, as it were, of the power of nature, which is to produce flowering and fruiting in later years.

IMAGE 9: Tiny seedling of lodgepole pine (*Pinus contorta*) exhibiting a circle of extended cotyledons

34

In several flowers we find unaltered stem leaves collected in a kind of calyx right under the flower. Since they retain their form clearly, we can rely on the mere appearance in this case, and on botanical terminology which calls them *folia floralia* (flower leaves).

IMAGE 10: Cornflower (*Centaurea montana*) with urn-shaped calyx beneath corolla

The Metamorphosis of Plants

IMAGE 11: "Thai Delight" *Bougainvillea* with leaflike bracts collected around the tubular flowers

35

We must now turn our attention to the instance mentioned above, where the transition to flowering occurs slowly as the stem leaves come together gradually, transform, and gently steal over, as it were, into the calyx. This can be observed quite clearly in the calyxes of the compositae, especially in sunflowers and calendulas.

IMAGE 12: Stem leaves and calyx of the sunflower (*Helianthus annuus*)

36

Nature's power to collect several leaves around one axis can create still closer connections, rendering these clustered, modified leaves even less recognizable, for it may merge them wholly or in part by making their edges grow together. The crowded and closely packed leaves touch one another everywhere in their tender state, anastomose through the influence of the highly purified juices now present in the plant, and produce a bell-shaped or (so-called) single-leaf calyx, which betrays its composite origins in its more or less deep incisions or divisions. We can see this if we compare a number of deeply incised calyxes with multileaved ones, and especially if we examine the calyxes of several compositae. Thus, for example, we will find that a calendula calyx (noted in systematic descriptions as *simple* and *much divided*) actually consists of many leaves grown into one another and over one another, with the additional intrusion, so to speak, of contracted stem leaves (as noted above).

IMAGE 13: Calyx of pot marigold (*Calendula officinalis*)

37

In many plants, the arrangement of individual or merged sepals around the axis of the stalk is constant in number and form; this is also true of the parts that follow. Biological science, which has developed significantly in recent years, has relied heavily on this consistency for its growth, stability, and reputation. The number and formation of these parts is not as constant in other plants, but even this inconsistency has not deceived the sharp eyes of the masters in this science; through exact definition they have sought to impose stricter limits, so to speak, on these aberrations of nature.

38

This, then, is how nature formed the calyx: it collected several leaves (and thus several nodes) around a central point, frequently in a set number and order; elsewhere on the plant these leaves and nodes would have been produced successively and at a distance from one another. If excessive nourishment had hampered flowering, they would have appeared in separate locations and in their original form. Thus, nature does not create a new organ in the calyx; it merely gathers and modifies the organs we are already familiar with, and thereby comes a step closer to its goal.

V. Formation of the Corolla

39

We have seen that the calyx is produced by refined juices created gradually in the plant itself. Now it is destined to serve as the organ of a further refinement. Even a simple mechanical explanation of its effect will convince us of this. For how delicate and suited for the finest filtration must be those tightly contracted and crowded vessels we have seen!

40

We can note the transition from the calyx to the corolla in several ways. Although the calyx is usually green like the stem leaves, the color of one or another of its parts often changes at the tip, edge, back, or even on the inner surface of a part where the outer surface remains green. We always find a refinement connected with this coloration. In this way, ambiguous calyxes arise that might equally well be called corollas (images 14,15).

41

In moving up from the seed leaves, we have observed that a great expansion and development occurs in the leaves, especially in their periphery; from here to the calyx, a contraction takes place in their circumference. Now we note that the corolla is produced by another expansion; the petals are usually larger than the sepals. The organs were contracted in the calyx, but now we find that the purer juices, filtered further through the calyx, produce petals that expand in a quite refined form to present us with new, highly differentiated organs. Their fine structure, color, and fragrance would make it impossible to recognize their origin, were we not able to get at nature's secret in several abnormal cases.

42

Within the calyx of a carnation, for example, there is often a second calyx: one part is quite green, with a tendency to form a single-leaf, incised calyx; another part is jagged, with tips and edges transformed into the delicate, expanded, colored, true beginnings of petals. Here we can again recognize the relationship between corolla and calyx.

43

The relationship between the corolla and the stem leaves is also shown in more than one way, for in several plants the stem leaves show some color long before the plant approaches flowering; others take on full coloration when flowering is near (image 16).

The Metamorphosis of Plants

IMAGE 14: Calyx and corolla of coreopsis (*Coreopsis grandiflora*) displaying both green and more refined golden sepals

IMAGE 15: Corolla of coreopsis (*Coreopsis grandiflora*)

IMAGE 16: Bee balm (*Monarda didyma*) showing advancing coloration in stem leaves. This individual also displays a second flower emerging from within the first.

44

Sometimes nature skips completely over the organ of the calyx, as it were, and goes directly to the corolla. We then have the opportunity to observe how stem leaves turn into petals. Thus, for example, an almost fully formed and colored petal often appears on tulip stems. It is even more remarkable when half of this leaf is green and attached as part of the stem, while its other, more colorful half rises up as part of the corolla, thereby dividing the leaf in two.

IMAGE 17: Transition from stem leaf to petal in the tulip (*Tulipa*)

45

It is probable that the color and fragrance of the petals are attributable to the presence of the male germ cell. Apparently it is still insufficiently differentiated in these petals, where it is combined and diluted with other juices. The beautiful appearance of the colors leads us to the notion that the material filling the petals has attained a high degree of purity, but not yet the highest degree (which would appear white and colorless).

VI. Formation of the Stamens

46

This becomes even more probable when we consider the close relationship between the petals and the stamens. Were the relationship between the other parts so striking, well known, and undeniable, there would be no need for this discourse.

47

Sometimes nature shows us this transition in an orderly way (e.g., in the canna and other plants of this family). A true petal, little changed, contracts at its upper border, and an anther appears, with the rest of the petal serving in place of the filament (image 18).

48

In flowers that frequently become double, we can observe every step of this transition. Within the fully formed and colored petals of several rose species there appear others that are partly contracted in the middle and partly at the side. This contraction is the result of a small thickened wale that somewhat resembles a perfect anther; the leaf likewise begins to assume the simpler form of a stamen (images 19, 20). In some double poppies, fully formed anthers rest on almost unaltered petals in the corolla (which is completely double); in others, the petals are more or less contracted by antherlike wales (images 20–24).

The Metamorphosis of Plants

IMAGE 18: *Canna x generalis* with anther arising from contracted petal

IMAGE 19 AND IMAGE 20: Full and contracted petals of a Damask rose (*Rosa damascena*), showing the relation of petals and stamens

The Metamorphosis of Plants

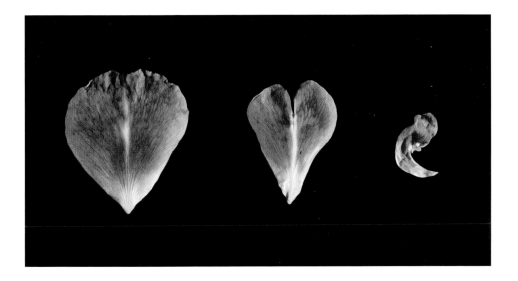

The Metamorphosis of Plants

IMAGE 21 AND IMAGE 22: Double poppy (*Papaver rhoeas*) displaying full petals

The Metamorphosis of Plants

IMAGE 23 AND IMAGE 24: Double poppy
(*Papaver atlantcium*) with partially contracted petals

The Metamorphosis of Plants

49

If all the stamens are transformed into petals, the flowers will be seedless; but if stamens develop even when a flower becomes double, fructification may occur.

50

Thus a stamen arises when the organs, which earlier expanded as petals, reappear in a highly contracted and refined state. This reaffirms the observation made above: we are made even more aware of the alternating effects of contraction and expansion by which nature finally attains its goal.

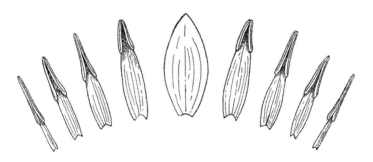

FIGURE 6: Successive transformation of petal into stamen in white water lily (*Nymphaea alba*)

VII. NECTARIES

51

However rapid the transition from corolla to stamens in many plants, we nonetheless find that nature cannot always achieve this in a single step. Instead, it produces intermediate agents that sometimes resemble the one part in form and purpose, and sometimes the other. Although they take on quite different forms, almost all may be subsumed under one concept: they are gradual transitions from the petals to the stamens.

52

Most of these variously formed organs (which Linnaeus[7] calls nectaries) may be subsumed under this concept. Here we are again bound to admire the intelligence of that extraordinary man: without any clear understanding of their purpose, he followed his intuition and ventured to use one name for such seemingly different organs.

53

Some petals show their relationship to the stamens without any perceptible change in form; they contain tiny cavities or glands that secrete a honeylike juice. In the light of our previous discussion, we may infer that this is an undeveloped and incompletely differentiated fluid of fertilization; our inference will be further justified in the discussion to follow.

54

The so-called nectaries may also appear as independent parts; these sometimes resemble the petals in form, and sometimes the stamens. Thus, for example, the thirteen filaments (each with a tiny red ball) on the nectaries of *Parnassia* have a striking resemblance to stamens (figures 7, 8). Other nectaries appear as stamens without anthers (as in *Vallisneria* or *Fevillea);* in *Pentapetes* we also find them, in leaf form, alternating with the stamens in a whorl; in addition, systematic descriptions describe them as *filamenta castrata petaliformia*.[8] We find equally unclear formations in *Kiggelaria* and the passion flower (image 25).

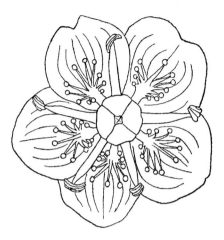

Figure 7: Flower of *Parnassia*, showing nectaries between stamens

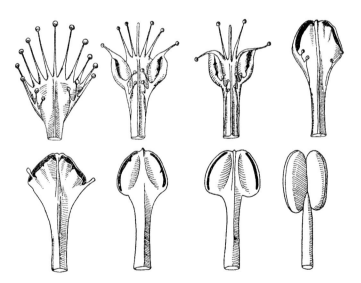

Figure 8: Intermediate forms of stamens and nectaries in *Parnassia*

The Metamorphosis of Plants

IMAGE 25: Passion flower (*Passiflora*) with its ambiguous nectaries

IMAGE 26: Primary and secondary corollas in *Narcissus*

55

The word *nectary* (in the sense indicated above) seems equally applicable to the distinctive secondary corolla. The formation of petals occurs by expansion, but secondary corollas are formed by contraction (that is, in the same way as the stamens). Within full, expanded corollas we therefore find small, contracted secondary corollas, as in the narcissus, *Nerium,* and *Agrostemma.*

56

We see even more striking and remarkable changes in the petals of other species. At the base of the petal in some flowers we find a small hollow filled with a honeylike juice. This little cavity is deeper in other species and types; it creates a projection shaped like a spur or horn on the back of the petal, thus producing an immediate modification in the form of the rest of the petal. We can observe this clearly in different types and varieties of the columbine.

The Metamorphosis of Plants

IMAGE 27: Columbine (*Aquilegia*), displaying spurred nectaries between petals

57

This organ is most transformed in the aconite and *Nigella*, for example, but even here its resemblance to the leaf is not hard to see. In *Nigella*, especially, it has a tendency to form again as a leaf, and the flower becomes double with the transformation of the nectaries. Careful examination of the aconite will show the similarity between the nectaries and the arched leaf under which they are hidden.

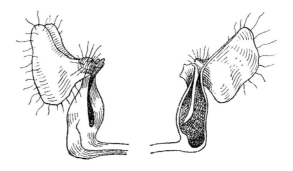

FIGURE 9: *Nigella damascena*, in entirety and in lengthwise section, with hollow depression and cover above it

IMAGE 28 (opposite): Monkshood (*Aconitum napellus*), with nectaries visible within "hoods"

FIGURE 10 (opposite): Original sketches by Goethe of various flower parts, among them nectaries of *Aconitum* and *Delphinium*

The Metamorphosis of Plants

58

We said above that the nectaries are transitional forms in the change from petal to stamen. Here we can make a few observations about irregular flowers. Thus, for example, the five outer leaves of *Melianthus* might be called true petals, but the five inner leaves could be described as a secondary corolla[9] consisting of six nectaries; the upper nectary is closest to the leaf in form, while the lower one (now called a nectary) is least like the leaf. In the same sense, we might say that the carina of the papilionaceous flowers is a nectary: of all the flower's leaves, it most resembles the stamens in form, and is quite unlike the leaf form of the so-called vexilla.[10] This also explains the brushlike appendages attached to the end of the carina in some species of *Polygala,* and thus it gives us a clear idea of the purpose these parts serve.

FIGURE 11: Flower of *Melianthus major L.* I Front view; II Side view; III Side view, calyx removed, the slipperlike nectary visible

The Metamorphosis of Plants

IMAGE 29: Papilionaceous (butterfly-shaped) flower of the sweet pea (*Lathyrus odoratus*), showing the curved, narrow carina at the center

FIGURE 12: Flower of *Polygala myrtifolia*

59

It should be unnecessary to state here that these remarks are not intended to confuse the distinctions and classifications made by earlier observers and taxonomists. Our only purpose is to help explain variations in plant form.

VIII. Further Remarks on the Stamens

60

Microscopic examination has shown beyond a doubt that the plant's reproductive organs are brought forth by spiral vessels,[11] as are the other organs. We will use this to support the argument that the different plant parts with their apparent variety of forms are nonetheless identical in their inner essence.

61

The spiral vessels lie amid the bundles of sap vessels, and are enclosed by them. We can better understand the strong force of contraction mentioned earlier if we think of the spiral vessels (which really seem like elastic springs) as extremely strong, so that they predominate over the expansive force of the sap vessels.

62

Now the shortened vessel bundles can no longer expand, join one another, or form a network by anastomosis; the tubular vessels that usually fill the interstices of the network can no longer develop, and there is nothing left to cause the expansion of stem leaves, sepals, and petals; thus a frail, very simple filament arises.

63

The fine membranes of the anther are barely formed, and the extremely delicate vessels terminate between them. Previously the vessels grew longer, expanded, and joined one another, but now we will assume that these same vessels are in a highly contracted state. We see a fully formed pollen emerge from them; in its activity this pollen replaces the expansive force taken from the vessels which produced it. Now released, it seeks out the female parts that the same effect of nature brings to meet it; it attaches itself to these parts, and suffuses them with its influence. Thus we are inclined to say that the union of the two genders is anastomosis on a spiritual level; we do so in the belief that, at least for a moment, this brings the concepts of growth and reproduction closer together.

64

The fine matter developed in the anthers looks like a powder, but these tiny grains of pollen are just vessels containing a highly refined juice. We therefore subscribe to the view that this juice is absorbed by the pistils to which the pollen grains cling, thereby causing fructification. This is made even more likely by the fact that some plants produce no pollen, but only a liquid.

IMAGE 30: Honeybee on Shasta daisy (*Leucanthemum x superbum*)

65

Here we recall the honeylike juice of the nectaries, and its probable relationship to the fully developed liquid of the pollen grains. Perhaps the nectaries prepare the way; perhaps their honeylike liquid is absorbed by the pollen grains, and then further differentiated and developed. This opinion is made more plausible by the fact that this juice can no longer be seen after fructification (image 31).

66

We will not forget to mention in passing that the filaments grow together in a variety of ways, as do the anthers. They offer the most wonderful examples of what we have often discussed: the anastomosis and union of plant parts that were, at first, strictly separate.

IX. Formation of the Style

67

Earlier I tried to make as clear as possible that the various plant parts developed in sequence are intrinsically identical despite their manifold differences in outer form. It should come as no surprise that I also intend to explain the structure of the female parts in the same way.

68

We will first examine the style apart from the fruit (as often found in nature). This will be all the easier since it is distinct from the fruit in this form.

69

We observe, then, that the style is at the same stage of growth as the stamens. We noted that the stamens are produced by a contraction; this is also true of the styles, and we find that they are either the same size as the stamens, or only a little longer or shorter in form. In many

The Metamorphosis of Plants

IMAGE 31: Trumpet-shaped flowers of orange honeysuckle
(*Lonicera ciliosa*), which hold nectar in their base

instances the style looks almost like a filament without anthers; the two resemble one another in external form more than any of the other parts. Since both are produced by spiral vessels, we can see plainly that the female part is no more a separate organ than the male part. When our observation has given us a clearer picture of the precise relationship between the female and male parts, we will find that the idea of calling their union an anastomosis becomes even more appropriate and instructive.

IMAGE 32: Flower of *Magnolia acuminate "Elizabeth"* with central style and surrounding stamens

IMAGE 33: Style and stamens in tiger lily (*Lilium lancifolium*)

70

We often find the style composed of several individual styles that have grown together; its parts are scarcely distinguishable at the tip, and sometimes not even separate. This is the most likely stage for this merger to occur; we have often mentioned its effects. Indeed, it must occur because the delicate, partially developed parts are crowded together in the center of the blossom, where they can coalesce.

71

In various cases of regular metamorphosis, nature gives a more or less clear indication of the close relationship between the style and the previous parts of the blossom. Thus, for instance, the pistil of the iris, with its stigma, appears in the full form of a flower leaf. The umbrella-shaped stigma of *Sarracenia* shows (although not so clearly) that it is composed of several leaves, and even the green color remains (image 35). With the aid of the microscope we will find the stigma of several flowers formed as full single-leaved or multi-leaved calyxes (for example, the crocus; or *Zannichellia*) (image 36).

72

In retrogressive metamorphosis nature frequently shows us instances where it changes the styles and stigmas back into flower leaves. *Ranunculus asiaticus,* for example, becomes double by transforming the stigmas and pistils of the fruit vessel into true petals, while the anthers just behind the corolla are often unchanged (figure 13). Several other noteworthy cases will be discussed later.

73

Here we will repeat our earlier observation that the style and the stamens are at the same stage of growth; this offers further evidence for the basic principle of alternation in expansion and contraction. We first noted an expansion from the seed to the fullest development

The Metamorphosis of Plants

IMAGE 34: Bearded iris (*Iris pallida*) with style (the "beard") and stigma in the form of a petal

IMAGE 35: Pitcher plant (*Sarracenia alata*), showing the umbrella-shaped stigma of the flower

The Metamorphosis of Plants

IMAGE 36: Multileaved stigma in *Crocus chrysanthus*

The Metamorphosis of Plants

Figure 13: Persian buttercup (*Ranunculus asiaticus*), with double flower

of the stem leaf; then we saw the calyx appear through a contraction, the flower leaves through an expansion, and the reproductive parts through a contraction. We will soon observe the greatest expansion in the fruit, and the greatest concentration in the seed. In these six steps nature steadfastly does its eternal work of propagating vegetation by two genders.

X. Of the Fruits

74

Now we come to the fruits. We will soon realize that these have the same origin as the other parts, and are subject to the same laws. Here we are actually speaking of the capsules formed by nature to enclose the so-called covered seeds, or, more precisely, to develop a small or large number of seeds by fructification within these capsules. It will not require much to show that these containers may also be explained through the nature and structure of the parts discussed earlier.

75

Retrogressive metamorphosis again makes us aware of this natural law. Thus, for example, in the pinks—these flowers known and loved for their irregularity—we often find that the seed capsules are changed back into leaves resembling those in the calyx, and the styles are accordingly shortened. There are even pinks in which the fruit capsule is completely transformed into a true calyx. The divisions at the tips of the calyx still bear delicate remnants of the styles and stigmas; a more or less full corolla develops instead of seeds from the very center of this second calyx.

IMAGE 37: Pinks (*Dianthus*), displaying retrogressive metamorphosis

76

Even in regular and constant formations, nature has many ways of revealing the fruitfulness hidden in a leaf. Thus an altered but still-recognizable leaf of the European linden produces a small stalk from its midrib, and grows a complete flower and fruit on this stalk. The disposition of blossoms and fruits on the leaves of *Ruscus* is even more remarkable (image 39).

IMAGE 38: Leaves of bigleaf linden (*Tilia platyphyllos*) showing midrib stalks

IMAGE 39: Butcher's broom (*Ruscus aculeatus*) with fruit

77

In the ferns we see still stronger—we might even say enormous—evidence of the sheer fruitfulness inherent in the stem leaves: these develop and scatter innumerable seeds (or rather, germs) through an inner impulse, and probably without any well-defined action by two genders. Here the fruitfulness of a single leaf rivals that of a wide-spreading plant, or even a large tree with its many branches.

IMAGE 40: Frond of autumn fern (*Dryopteris erythrosora*) displaying red spore cases

78

With these observations in mind, we will not fail to recognize the leaf form in seed vessels—regardless of their manifold formations, their particular purpose and context. Thus, for example, the pod may be viewed as a single, folded leaf with its edges grown together; husks, as consisting of leaves grown more over one another; and compound capsules may be understood as several leaves united round a central point, with their inner sides open toward one another and their edges joined (images 41, 42). We can see this for ourselves when these compound capsules burst apart after maturation, for each part will then present itself as an open pod or husk. We may also observe a similar process taking place regularly in different species of the same genus: the fruit capsules of *Nigella orientalis*, for instance, are partially merged pods grouped around an axis; but in *Nigella damascena* they are fully merged (figures 14, 15).

79

Nature masks the resemblance to the leaf mainly by forming soft, juicy seed vessels, or hard, woody ones. But this similarity will not escape our attention if we know how to follow it carefully through all its transitions. Here we will have to be content with having given a description of the general concept along with several examples of nature's consistent behavior. The great variety in seed capsules will provide material for a great many other observations in the future.

80

The relationship between the seed capsules and the previous parts also appears in the stigma, situated right on top of the seed capsule and inseparably joined to it. We have already demonstrated the relationship of the stigma to the leaf form, and here we may note it again: in double poppies we find that the stigmas of the seed capsules are changed into delicate, colored leaflets that look exactly like petals.

The Metamorphosis of Plants

IMAGE 41: Seedpods of Oriental poppy (*Papaver orientale*)

IMAGE 42: Seedpods of *Iris* with outer covering partially removed to reveal seeds

The Metamorphosis of Plants

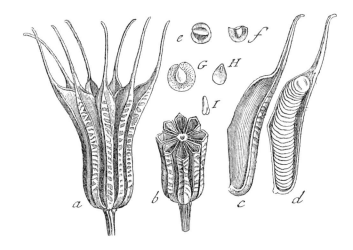

Figure 14: Fruit capsules of *Nigella orientalis*

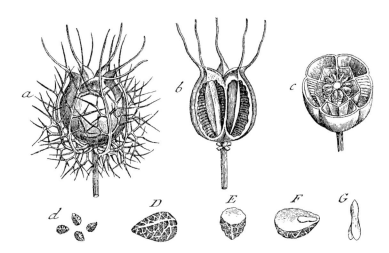

Figure 15: Fruit capsules of *Nigella damascena*

81

The last and most pronounced expansion in the growth of the plant appears in the fruit. This expansion is often very great—even enormous—in inner force as well as outer form. Since it usually occurs after fertilization, it seems likely that as the developing seed draws juices from the entire plant for its growth, the flow of these juices is directed into the seed capsule. The vessels of the seed capsule are thereby nourished and expanded, often becoming extremely gorged and swollen. It can be inferred from our earlier discussion that purer gases play a part in this, an inference supported by the discovery that the distended pods of *Colutea* contain a pure gas.[12]

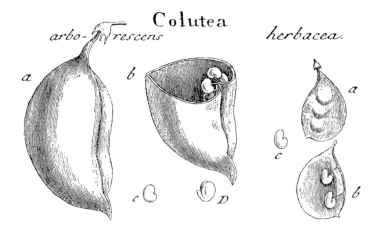

FIGURE 16: Hulls of *Colutea arborescens* and *Colutea herbacea*, showing their leaflike character

XI. Of the Coverings Lying Next to the Seed

82

By way of contrast, the seed is in the most extreme state of contraction and inner development. In various plants we can observe that the seed transforms leaves into an outer covering, adapts them more or less to its shape, and often has the power to annex them fully, completely changing their form. We saw above that many seeds can develop in and from a single leaf; hence it will come as no surprise to find a single embryo clothed in a leaf covering.

83

We can see the traces of such incompletely adapted leaf forms in many winged seeds (e.g., the maple, the elm, the ash, and the birch) (images 43–46). The calendula's three distinct rings of differently formed seeds offer a remarkable example of how the embryo pulls broad coverings together, gradually adapting them to its shape (figure 17). The outer ring is still related to the petals in form, except that a rudimentary seed swells the rib, causing a fold in the leaf; a small membrane also runs lengthwise along the inside of the crease, dividing the leaf in two. The next ring shows further changes: the broad form of the leaf has entirely disappeared, along with the membrane; but its shape is somewhat less elongated, while the rudimentary seed on the back has become more visible, and the small raised spots on the seed have grown more distinct. These two rows appear to be either unfructified or only partially fructified. They are followed by a third row of seeds in their true form: strongly curved, and with a tightly tilted involucre that is fully developed in all its ridges and raised portions. Here we again see a powerful contraction of broad, leaflike parts, a contraction produced by the inner power of the seed, just as we earlier saw the flower leaf contracted by the power of the anthers.

IMAGE 43: Leaves and winged seeds of Japanese maple (*Acer palmatum*)

IMAGE 44: Leaves and seeds of Camperdown elm (*Ulmus glabra*)

IMAGE 45: Leaves and seeds of green ash (*Fraxinus pennsylvanica*)

IMAGE 46: Leaves and seed-bearing catkins of paper birch (*Betula papyrifera*)

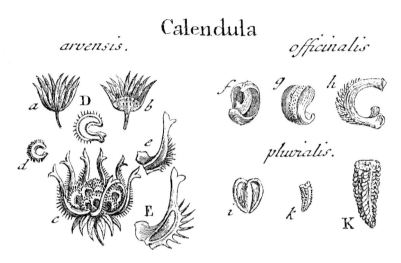

Figure 17: Seeds of *Calendula*

XII. Review and Transition

84

Thus we have sought to follow as carefully as possible in the footsteps of nature. We have accompanied the outer form of the plant through all its transformations, from the seed to the formation of a new seed; we have investigated the outer expression of the forces by which the plant gradually transforms one and the same organ, but without any pretense of uncovering the basic impulses behind the natural phenomena. So as not to lose the thread that guides us, we have limited our discussion entirely to annual plants; we have noted only the transformation of the leaves accompanying the nodes, and have derived all the forms from them. But to lend our discussion the required thoroughness, we must now speak of the eyes hidden beneath each leaf; under certain circumstances these develop, and under others they seem to disappear entirely.

XIII. Of the Eyes and Their Development

85

Nature has given each node the power to produce one or more eyes; this process takes place near its companion leaves, which seem to prepare the way for the formation and growth of the eyes, and help in their production.

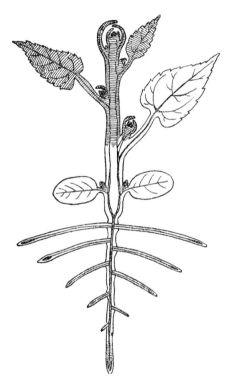

Figure 18: Diagrammatic sketch of young dicotyledonous plant: white = parts already developed; hatched = those still in process of extension and growth; black = youngest parts. In axils of cotyledons and leaves proper, the "eyes," or buds, can be seen.

86

The primary, simple, slow process of plant reproduction is based on the successive development of one node from the other, and the growth of an eye close to it.

87

We know that such an eye is similar to the ripe seed in its effect; in fact, we can often recognize the whole shape of the potential plant more easily in the eye than in the seed.

88

Although the root point[13] is hard to find in the eye, it is just as much there as in the seed, and will develop quickly and easily, especially in the presence of moisture.

IMAGE 47: Potato (*Solanum tuberosum*) with sprouts growing from eyes

89

The eye needs no cotyledon because it is connected to the fully developed parent plant, and receives adequate nourishment as long as the connection remains. Once separated, it will draw nourishment from the plant to which it is grafted, or from the roots developed as soon as a branch is planted in the earth.

90

The eye consists of more or less developed nodes and leaves that have the task of enhancing the future growth of the plant. Thus the side branches growing from the nodes of the plant may be considered separate small plants placed on the parent in the same way that the parent is attached to the earth.

91

The two have often been compared and contrasted, most recently in such an intelligent and exact way that we will simply refer to it here with our unqualified admiration (Gaertner, *De fructibus et seminibus plantarum,* Chapt. I)[14]

92

We will say only the following on this point. Nature makes a clear distinction between eyes and seeds in plants with a highly differentiated structure. But if we descend to plants with a less differentiated structure, the two become indistinguishable, even for the sharpest observer. There are seeds that are clearly seeds, and gemmae that are clearly gemmae,[15] but it takes an act of reason rather than observation to find the connection between the seeds, which are actually fertilized and separated from the parent plant by the reproductive process, and the gemmae, which simply grow out of the plant and detach without apparent cause.

93

With this in mind, we may conclude that the seeds are closely related to the eyes and gemmae, although they differ from the eyes in being enclosed, and from the gemmae in having a perceptible cause for their formation and separation.

XIV. Formation of Composite Flowers and Fruits

94

Thus far we have focused on the transformation of nodal leaves in our attempt to explain the development of simple flowers, as well as the production of seeds enclosed in capsules. Closer examination will show that no eyes form in these cases, and moreover, that the formation of such eyes is utterly impossible. We must look to the formation of eyes, however, to explain the development of composite flowers or compound fruit arranged around a single cone, a single spindle, a single disk, etc. (images 48–50).

95

Certain stems do not gradually prepare the way for a single flower by saving their energies; instead, they produce their flowers directly from the nodes, and frequently continue this process without interruption to their very tip. This phenomenon may he explained, however, through the theory presented earlier. All flowers developed from the eyes must be considered whole plants situated on the parent, just as the parent is situated on the earth. Since they now receive purer juices from the nodes, even the first leaves of the tiny twig appear much more fully developed than the first leaves (following the cotyledons) of the parent; in fact, it is often possible to develop the calyx and flower immediately.

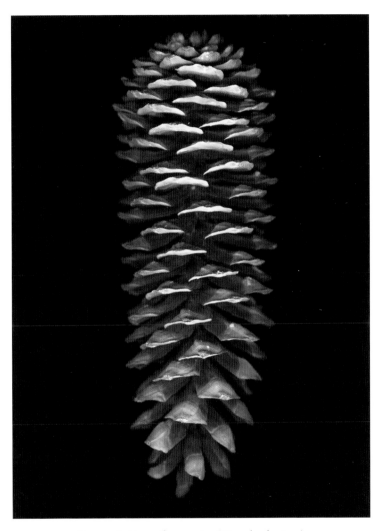

IMAGE 48: Cone of sugar pine (*Pinus lambertiana*)

IMAGE 49: Spindle-shaped composite flower of the butterfly bush (*Buddleja davidii*)

The Metamorphosis of Plants

IMAGE 50: Gerbera daisy (*Gerbera jamesonii*), showing ray florets surrounding the circular area of disk florets. In the plant on the right, a number of ray florets have metamorphosed from the central disk.

96

With an increase in nourishment, the flowers developed from the eyes would become twigs; they are necessarily subject to the same conditions as the parent stem, and share in its fate.

97

As these flowers develop from node to node, we also find that the stem leaves undergo the same changes seen previously in the gradual transition to the calyx. They contract more and more, finally disappearing almost completely, and they are called bracts when their form has become somewhat different from a leaf. The stem likewise grows thinner, the nodes crowd closer together, and all the phenomena noted earlier take place, but there is no decisive formation of a flower at the end of the stem because nature has already exercised its rights from node to node.

98

Having examined the stem adorned with a flower at every node, we will soon arrive at an explanation of the *composite flower,* especially if we recall what was said before about the creation of the calyx.

99

Nature forms a composite calyx out of many leaves compacted around a single axis. Driven by the same strong growth impulse, it suddenly develops an endless stem, so to speak, with all its eyes in the form of flowers and compacted as much as possible; each small flower fertilizes the seed vessel standing ready below. The nodal leaves are not always lost in this enormous contraction; in the thistles, the little leaves faithfully accompany the floret developed from the eye next to them (compare the form of *Dipsacus laciniatus*). In many grasses, each flower is accompanied by such a little leaf (called a glume).

The Metamorphosis of Plants

IMAGE 51: Thistle (*Cirsium edule*) with blossoms at nodes and accompanying nodal leaves

IMAGE 52: Stem of common teasel (*Dipsacus*), displaying dried flower heads and upcurved bracts

IMAGE 53: Spike of Italian ryegrass (*Lolium multiflorum*), showing pointed glumes at the base of each floret

100

Thus we now realize that the seeds developed around a composite flower are true eyes created and formed by the reproductive process. With this concept firmly in mind, we may compare a variety of plants, their growth and their fruits, and find convincing evidence in what we see.

101

Hence, it will not be hard to explain the covered or uncovered seeds produced in the center of a single flower, often in a group around a spindle. For it is all the same, whether a single flower surrounds a common ovary where the merged pistils absorb the reproductive juices from the flower's anthers and infuse them into the ovules, or whether each ovule has its own pistil, its own anthers, and its own petals around it.

102

We are convinced that with a little practice the observer will find it easy to explain the various forms of flowers and fruits in this way. To do so, however, requires that he feel as comfortable working with the principles established above—expansion and contraction, compaction and anastomosis—as he would with algebraic formulas. Here it is crucial that we thoroughly observe and compare the different stages nature goes through in the formation of genera, species, and varieties, as well as in the growth of each individual plant. For this reason alone, it would be both pleasant and useful to have a collection of properly arranged illustrations labeled with the botanical terms for the different parts of the plant. In connection with the above theory, two kinds of proliferous flowers would serve as especially useful illustrations.

XV. Proliferous Rose

103

The proliferous rose offers a very clear example of everything we sought earlier through our power of imagination and understanding. The calyx and corolla are arranged and developed around the axis, but the seed vessel is not contracted in the center with the male and female organs arranged around it. Instead, the stem, half reddish and half greenish, continues to grow, developing a succession of small, dark red, folded petals, some of which bear traces of anthers. The stem grows further; thorns reappear on it; one by one, the colored leaves that follow become smaller; and finally we see them turn into stem leaves, partly red and partly green. A series of regular nodes forms, and from their eyes small but imperfect rosebuds once again appear (figure 19).

104

This example also gives visible evidence of another point made earlier; that is, that all calyxes are only contracted *folia floralia*.[16] Here the regular calyx gathered around the axis consists of five fully developed, compound leaves with three or five leaflets, the same sort of leaf usually produced by rose branches at their nodes.

FIGURE 19: Proliferous rose. Watercolor commissioned by Goethe in the early 1790s, intended for a sequel to *The Metamorphosis of Plants*

XVI. Proliferous Carnation

105

After spending some time with this phenomenon, we may turn to another that is still more remarkable: the proliferous carnation. We see a perfect flower equipped with a calyx as well as a double corolla and completed in the center with a seed capsule, although this is not fully developed. Four perfect new flowers develop from the sides of the corolla; these are separated from the parent flower by stalks having three or more nodes. They have their own calyxes, and double corollas formed not so much by individual leaves as by leaf crowns merged at the base, or more often by flower leaves that have grown together like little twigs around a stem. Despite this extreme development, filaments and anthers are found in some. We see fruit capsules with styles, and seed receptacles that have grown back into leaves; in one such flower the seed envelopes had joined to create a full calyx containing the rudiments of another perfect double flower.

106

In the rose we have seen a partially defined flower, as it were, with a stem growing again from its center, and new leaves developing on this stem. But in this carnation, with its well-formed calyx, perfect corolla, and true seed capsules in the center, we find that eyes develop from the circle of petals, producing real branches and blossoms. Thus both instances illustrate that nature usually stops the growth process at the flower and closes the account there, so to speak; nature precludes the possibility of growth in endless stages, for it wants to hasten toward its goal by forming seeds (figure 20).

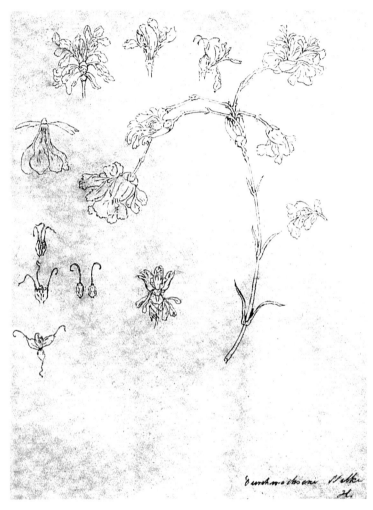

Figure 20: Proliferous carnation. Original sketch by Goethe probably done in 1787

XVII. Linnaeus's Theory of Anticipation

107

If I have stumbled here and there on the path that a predecessor described as terrifying and dangerous, even though he attempted it under the guidance of his great teacher (Ferber, *Diss. de prolepsi plantarum*);[17] if I have not done enough to pave the way for those who follow; if I have not cleared every obstacle from the path—nonetheless, I hope that this effort will not prove altogether fruitless.

108

It is now time to consider a theory proposed by Linnaeus to explain these phenomena.[18] The things discussed here could not have escaped his sharp eyes; if we have made progress where he faltered, it is only because of a concerted effort by other observers and thinkers to clear the way and eliminate prejudice. A full comparison between his theory and the above discussion would be too time consuming here. The knowledgeable reader can make the comparison himself, but it would require too much detailed explanation to clarify it here for those who have not yet studied these things.

109

He started with an observation of trees, those complex and long-lived plants. He observed that a tree planted in a wide pot and over-fertilized would produce branch after branch for several years, while the same tree in a smaller pot would quickly bear blossoms and fruits. He saw that the successive development of the first tree was suddenly compressed in the second. He called this effect of nature *prolepsis* (anticipation) since the plant seemed to anticipate six years' growth in the six steps noted above.[19] He therefore developed his theory from tree buds; he did not pay much attention to annual plants, for he could see that these did not fit his theory as well. His theory would have us assume that nature really intended every annual plant to grow for six years, but the plant forestalled this maturation period by quickly blossoming, bearing fruit, and then dying.

110

We, however, began by following the growth of annual plants. Our approach is readily applicable to longer-lived plants, for a bud opening on the oldest tree may be considered an annual plant even though it develops on a long-existent stem and may itself last for a longer time.

111

There was a second reason for Linnaeus's lack of progress: he mistakenly viewed the various concentric parts of the plant (the outer bark,[20] the inner bark, the wood, the pith) as similar in their effect, similar in the way they participated in the life of the plant. He identified the various rings of the stem as the source of blossom and fruit because the latter, like the former, enclose one another and develop out of one another. But this was merely a superficial observation that closer examination shows to be false. The outer bark is unsuited to yield anything further; in the long-lived tree it is too separate and too hardened on the outside, just as the wood becomes too hard on the inside. In many trees the outer bark drops away, and in others it can be peeled without causing damage; thus it produces neither calyx nor any other living part of the tree. It is the second bark that contains all the power of life and growth; to the extent it is damaged, the tree's growth is also hindered. After examining all the external parts of the tree, we will discover that this is the part that brings growth gradually in the stem, and quickly in the flower and fruit. Linnaeus assigned it the mere secondary task of producing petals. By contrast, he assigned to the wood the important job of producing stamens, although we can see that the wood is rendered inactive by its solidity; it is durable but too dead to produce life. He supposed the pith to have the most important function: production of the pistils and numerous offspring. Yet doubts about the great importance of the pith seem to me significant and conclusive, as do the reasons for raising them.[21] The style and fruit merely appear to develop from the pith because our first impression is of soft, ill-defined, pithlike, parenchymatous formations gathered together in the center of the stem where we usually see only the pith.

XVIII. Recapitulation

112

I hope that this attempt to explain the metamorphosis of plants may contribute something to the resolution of these doubts, and lead to further findings and conclusions. The observations that serve as the basis for my work were made at various times, and have already been collected and organized (Batsch, *Introduction to the Identification and History of Plants,* Part I, Chapt. 19).[22] It should not be long before we discover whether the step taken here brings us any closer to the truth. We will summarize the principal results of the foregoing treatise as briefly as possible.

113

If we consider the plant in terms of how it expresses its vitality, we will discover that this occurs in two ways: first through growth (production of stem and leaves); and secondly, through reproduction (culminating in the formation of flower and fruit). If we examine this growth more closely, we will find that as the plant continues from node to node, growing vegetatively from leaf to leaf, a kind of reproduction also takes place, but a reproduction unlike that of flower and fruit; whereas the latter occurs all at once, the former is successive and appears as a sequence of individual developments. The power shown in gradual vegetative growth is closely related to the power suddenly displayed in major reproduction. Under certain circumstances a plant can be made to continue its vegetative growth, and under others the production of flowers can be forced. The former occurs when cruder juices accumulate; the latter, when more rarefied juices predominate.

114

In saying that vegetative growth is successive reproduction, while flowering and fruiting are simultaneous reproduction, we are also describing how each occurs. A vegetating plant expands to some extent, developing a stalk or stem; the intervals between nodes are usually perceptible, and its leaves spread out on all sides. A blossoming plant, on the other hand, shows a contraction of all its parts; the dimensions of length and breadth are canceled out, as it were; all its organs develop in a highly concentrated state and lie next to one another.

115

Whether the plant grows vegetatively, or flowers and bears fruit, the same organs fulfill nature's laws throughout, although with different functions and often under different guises. The organ that expanded on the stem as a leaf, assuming a variety of forms, is the same organ that now contracts in the calyx, expands again in the petal, contracts in the reproductive apparatus, only to expand finally as the fruit.

116

This effect of nature is accompanied by another: the gathering of different organs in set numbers and proportions around a common center. Under certain conditions, however, some flowers far exceed these proportions, or vary them in other ways.

IMAGE 54: Vegetative and reproductive organs displayed in "Coral Nymph" salvia (*Salvia coccinea*)

117

Anastomosis also plays a part in the formation of flowers and fruits; the extremely crowded and delicate organs of fructification are merged during the whole of their existence, or at least some part of it.

118

The phenomena of convergence, centering, and anastomosis are not peculiar to flower and fruit alone. We can discover something similar in the cotyledons, and ample material will be found in other parts of the plant for further observations of this sort.

119

We have sought to derive the apparently different organs of the vegetating and flowering plant from one organ; that is, the leaf normally developed at each node. We have likewise ventured to find in the leaf form a source for the fruits that completely cover their seed.

120

Here we would obviously need a general term to describe this organ that metamorphosed into such a variety of forms, a term descriptive of the standard against which to compare the various manifestations of its form.[23] For the present, however, we must be satisfied with learning to relate these manifestations both forward and backward. Thus we can say that a stamen is a contracted petal or, with equal justification, that a petal is a stamen in a state of expansion; that a sepal is a contracted stem leaf with a certain degree of refinement, or that a stem leaf is a sepal expanded by an influx of cruder juices.

121

We might likewise say of the stem that it is an expanded flower and fruit, just as we assumed that the flower and fruit are a contracted stem.

122

At the conclusion of the treatise I also took the development of eyes into account, and attempted thereby to explain composite flowers as well as uncovered fruits.

123

Thus I have tried to be as clear and thorough as I could in presenting a view I find rather convincing. Nonetheless, the evidence may still seem insufficient, objections may still arise, and my explanations may sometimes not seem pertinent. I will be all the more careful to note any suggestions in the future, and will discuss this material in a more precise and detailed way so that my point of view becomes clearer; perhaps then it will be more deserving of applause than at present.

Notes

1. A reference to plant reproduction.
2. Eighteenth-century botany used the term *vessel* for the various anatomical components of the leaves.
3. *Anastomosis* refers to the union of separate parts into an integrated network or whole.
4. Johannes Hedwig (1730–1799), physician and director of the Leipzig Botanical Gardens; in his article (1781) he opposed Linnaeus's explanation of the reproductive role of the pith.
5. Goethe's explanation of metamorphosis through the refinement of juices is based in the medical work of Hippocrates and Paracelsus.
6. The *organs of the nodes* are the leaves.
7. Carl von Linne (1707–1778), Swedish botanist, the founder of modern systematic classification in botany.
8. Emasculated filaments.
9. The corona.
10. The *carina* is the "keel" formed in papilionaceous flowers by the merger of two petals; papilionaceous flowers have the shape of a butterfly (typical of leguminous plants). The *vexilla* are the large "standards" (petals) found at the back of some flowers (for example, the pea).

11. Goethe's concept of these vessels is based on Hedwig's work (see note 4).
12. Goethe had chemists analyze gases from the pods of *Colutea arborescens*, a leguminous shrub with bladderlike pods.
13. The *root point* is the point (found in both seed and eye) where the plant will begin growing down into the ground rather than up into the air. In a note on plant metamorphosis from 1819, Goethe writes, "The seed already contains the organs which will divide the plant into two parts: those with a decided tendency to grow into the earth, seeking moisture and darkness, and those with a need to grow up into the light and air. When such a point manifests itself, we can think of it as located at or just below the surface of the earth. . . . As I picture it, the point which divides the stem from the root actually has the character of an ideal point, and cannot be counted as the first node."
14. Joseph Gaertner (1732–1791), German botanist, published the two main volumes of *De fructibus et seminibus plantarum*, a seminal study of fruits and seeds, in 1788 and 1791.
15. The *gemmae* are propagative buds.
16. On *folia floralia* see para. 34.
17. Johann Jacob Ferber (1743–1790), a student of Linnaeus and author of *Diss. de prolepsi plantarum* (1763).
18. Linnaeus's theory of anticipation was presented in *Prolepsis plantarum* (1760).
19. See para. 73.
20. As used by Goethe, the term *bark* includes the epidermis, bark, cortex, phloem, and cambium.
21. See note 4.
22. August Johann Georg Carl Batsch (1761–1802), German botanist, professor of medicine and natural history at the University of Jena, and botanical advisor to Goethe, published the book cited in 1787/88.
23. A reference to the "archetypal plant" for which Goethe had been searching during his Italian journey. In May, 1787, he wrote to Herder from Naples: "The primal plant is going to be the strangest creature in the world, which nature herself shall envy me. With this model and the key to it, it will be possible to go on forever inventing plants and know that their existence is logical; that is to say, if they do not actually exist, they could, for they are not the shadowy phantoms of a vain imagination, but possess an inner necessity and truth. The same law will be applicable to all other living organisms." (*Italian Journey*, London: Collins/Penguin, 1962, 310–311).

Appendix
The Genetic Method

Goethe was deeply interested in scientific method, realizing as he did that the answers one gets from inquiries into nature depend to a large extent on how one poses the questions. In his botanical work, he was of course concerned primarily with the "how" of vegetation and therefore investigated not only the diversity of physical forms but also the underlying unity from which they emerge. In a sketch of his distinctive approach to this type of investigation, written in the mid-1790s, he presents what he calls the "genetic method." The term *genetic* here refers not to the science of genes, but rather to seeking the origin or genesis of things. He describes this method as follows:

> If I look at the created object, inquire into its creation, and follow this process back as far as I can, I will find a series of steps. Since these are not actually seen together before me, I must visualize them in my memory so that they form a certain ideal whole.
>
> At first I will tend to think in terms of steps, but nature leaves no gaps, and thus, in the end, I will have to see this progression of uninterrupted activity as a whole. I can do so by dissolving the particular without destroying the impression itself.[1]

Appendix

Goethe believed that by practicing this method one could learn to consciously move back and forth between the region of relatively fixed and finished forms to the deeper realm of formative process. In the spirit of Spinoza, he was proposing that nature can be conceived in two ways—as creative power and as created product, or, in Spinoza's terminology, as *Natura naturans* ("nature naturing") and as *Natura naturata* ("nature natured"). And he worked to complement empiricism with imagination in order to see nature complete and unified as both creator and creation. As he suggests in his poem on the metamorphosis of plants, when faced with the daunting profusion of botanical forms, the way to uncover the simplicity of the "secret law" is to "gaze on them as they grow."

How does the genetic method work in practice? While Goethe saw this method as applicable to the overall metamorphosis of a plant, it is easier to see in a subset of that larger process—the sequence of changing leaf forms sequentially displayed on the stems of many plants. The images included here show six such leaf sequences, depicted in the order in which the leaves appear together on the stem. Consider the leaves of *Sidalcea malviflora*. The four leaves represent steps in the metamorphic process. To descend via these diverse steps to the implicit wholeness at their source, one needs first to give close attention to the particular forms themselves. Beginning with the bottommost leaf on the right, we study its features intently—visually inspecting the rounded shape, the relative regularity of the scalloped edge, and the structure of the veins. Moving to the next leaf, on the left, we find the overall roundness somewhat modified by the appearance of small incisions that were only hinted at in the previous leaf. Proceeding on to the third and fourth leaves, we can see the formative process articulated more fully. The leaves become larger and less rounded, the incisions grow into definite divisions, but the original plan is still evident in the pattern of the veins. Thus there is a sameness in the midst of the differences.

Appendix

IMAGE 55: Leaf sequence in *Sidalcea malviflora*

While in the case of *Sidalcea* we saw the leaf forms move from simple to complex, with *Delphinium astolat* the ascending movement is from complex to simple. We can nevertheless see evidence of the same formative principles working, though in different directions, in these two illustrations, as we can as well in the remaining four specimens.

Botanist Jochen Bockemühl has identified four basic movements in the spatial, archetypal dynamics of leaf formation—stemming, spreading, articulating, and shooting, as shown in figure 21. Together these movements constitute a logic of development in the metamorphosis of leaves, with the forces of intensification and polarity evident throughout.[2]

Looking past the leaf sequences depicted here to the overall metamorphosis of these plants, the vegetative leafy phase soon gives way to the reproductive, with varying degrees of contraction coming into play as the process moves from the stem leaves to the calyx and beyond.

The second part of the genetic method requires what Goethe called "exact sensory imagination." We initially see the different leaves as discrete steps in a process, but since "nature leaves no gaps," we need to consolidate these steps in order to apprehend nature's continuous inner workings. Reviewing the sequence of leaves, we then attentively internalize these visual forms as memory images. With these forms firmly in mind, we move in imagination through the sequence, transforming the first into the second, the second into the third, and so on, following the process forward and backward, forward and backward, as nature has also done. We thus implicate each explicit form—each momentary pause in the process—with those before and after, like the flow of notes in a musical performance. By focusing on the relationship between the leaf forms, exact sensory imagination involves setting one's mind in corresponding

Appendix

IMAGE 56: Leaf sequence in *Delphinium astolat*

Appendix

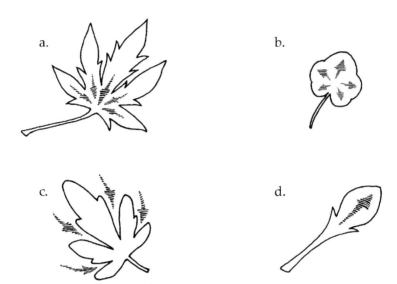

Figure 21: Dynamics of leaf formation (a) stemming, (b) spreading, (c) articulating, (d) shooting

motion, so that the selfsame living idea that has expressed itself in the metamorphosis of the plant comes to life and visibility in the mind as well. What was successive in one's empirical experience then becomes simultaneous in the intuitively perceived idea—*Proteus in potentia*. Instead of an onlooking subject knowing an alien object, this is knowledge through participation, or even identification, of observer and observed—knowing things from the inside. As Goethe said, "our spirit stands in harmony with those simpler powers that lie deep within nature; and it is able to represent them to itself just as purely as the objects of the visible world are formed in a clear eye."[3]

Goethe thought that moving from fixed forms to formative process—from parts to whole—requires shifting mental gears. He called the two cognitive faculties involved in this effort "understanding," which is the rational thinking that is the common instrument of conventional science, and "reason," the intuitive perception that sustains the poetic sensibility. Both of these mental modes play important roles in science and in life, but they do not provide equal access to the heart, or mind, of nature:

> The Understanding will not reach her; man must be capable of elevating himself to the highest Reason, to come into contact with the Divinity, which manifests itself in the primitive phenomena (*Urphänomenen*), which dwells behind them, and from which they proceed.
>
> The Divinity works in the living, not in the dead; in the becoming and changing, not in the become and the fixed. Therefore Reason, with its tendency toward the Divine, has only to do with the becoming, the living; but Understanding with the become, the already fixed, that it may make use of it.[4]

The genetic method encompasses both understanding and reason, attempting to unite the two for their mutual benefit—joining science and poetry—and Goethe criticized the one-sided emphasis of contemporary science on understanding alone, which served to limit its inquiries to merely the material surfaces of the natural world. He derived these two terms from Immanuel Kant's *Verstand* and *Vernunft* respectively. Kant, however, who influenced Goethe in many ways, felt that intuitive perception—*Vernunft*—was impossible to achieve. Goethe, on the other hand, boldly believed that "through an intuitive perception of eternally creative nature we may become worthy of participating spiritually in its creative processes."[5] The genetic method holds out the hope not only of revealing some deeper secrets of nature but also of releasing new powers of mind. Germany's greatest poet-scientist therefore, was fully aware, and unabashedly hopeful, that perceiving the essence of metamorphosis will likely involve a beneficial metamorphosis in the essence of the perceiver.

Notes

1. Goethe, "Studies for a Physiology of Plants," in *Scientific Studies*, edited and translated by Douglas Miller (New York: Suhrkamp, 1988), 75.
2. Jochen Bockemühl, "Transformations in the Foliage Leaves of Higher Plants," in *Goethe's Way of Science*, edited by David Seamon and Arthur Zajonc (Albany: SUNY Press, 1998), 115–128.
3. Goethe, quoted in Richards, *The Romantic Conception of Life*, (Chicago: University of Chicago Press, 2002), 377 n. 152.
4. Goethe, *Conversations with Eckermann* (San Francisco: North Point Press, 1984), 238 (13 Feb. 1829).
5. Goethe, "Judgement through Intuitive Perception," in *Scientific Studies*, 31.

Appendix

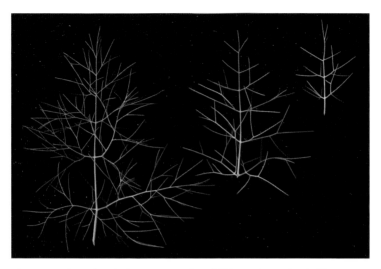

IMAGE 57: Ascending leaf structure in fennel (*Foeniculum vulgare*). This is one of the leaf sequences Goethe became fascinated with during his Italian journey.

IMAGE 58: Sequence of ascending leaf forms in wall lettuce (*Lactuca muralis*), from left to right

Appendix

IMAGE 59: Ascending leaf sequence in *Scabiosa columbaria*

Appendix

IMAGE 60: Leaf sequence in creeping buttercup *(Ranunculus repens)*

Sources

Epigraph from Rudolf Steiner, *Nature's Open Secret: Introductions to Goethe's Scientific Writings,* translated by John Barnes and Mado Spiegler (Great Barrington, Mass.: Anthroposophic Press, 2000).

Text of *The Metamorphosis of Plants* reprinted from *Goethe's Collected Works,* Vol. 12, *Scientific Studies,* edited and translated by Douglas Miller (New York: Suhrkamp, 1988), by permission of the publisher.

Text of "The Metamorphosis of Plants" (Goethe poem) reprinted from Rudolf Magnus, *Goethe as a Scientist,* translated by Heinz Norden (New York: Henry Schuman, 1949).

Figure A: portrait of Goethe by C. A. Schwerdgeburth, reprinted by permission of Smithsonian Institution Libraries.

Figures reprinted by permission of the Goethe-und Schiller-Archiv are as follows: figure B (GSA 26/LIV, 8); figure C (GSA 26/LIV, 8); figure 5 (GSA 26/LXI, 3, 16 Bl 153); figure 10 (GSA 26/LXI, 3, 22 Bl 156); figure 19 (GSA 26/LIV, 8); figure 20 (Goethe-Nationalmuseum unserer Klassik Stiftung Weimar).

Figures 1, 2, 3, 4, 6, 7, 8, 9, 11, 12, 13, 14, 15, 16, 17, and 18 reprinted from Wilhelm Troll, *Goethes Morphologische Schriften* (Jena: Eugen Diederichs, 1932).

Captions for figures 2, 4, 6, 7, 8, 9, 11, 16, and 18 reprinted from *Goethe's Botanical Writings,* translated by Bertha Mueller (Honolulu: University Press of Hawaii, 1952), by permission of the publisher.

Index

Page numbers in italics indicate images and figures.

Abnormalities, xxvi, 5, 6–7, 30, 31, 52
Acer palmatum, *76*
Aconitum napellus, 50, *51*
Agrostemma, 48
Air (effects on metamorphosis), 19, *21*
Altitude (effects on metamorphosis), 19, *20*
Anastomosis, 19, 28, 54, 55, 56, 57, 92, 102, 103n3
Annual plants (Goethe's focus on), 6, *7*, 80, 97–98
Anticipation (Linnaeus's theory of), 97–99, 104n18
Aquilegia, 48, *49*
Archetype, xvii–xviii, xix, xxii, xxiii, xxiv, 104n23. *See also* Urpflanze
Ash, 75, *78*

Bark, 22, 98, 104n20
Batsch, August Johann Georg Carl, 99, 104n22
Bean, *11,* 12, *12, 13*
Bee balm, *34*
Betula papyrifera, *79*
Birch, 75, *79*
Bockemühl, Jochen, 108
Bougainvillea, *27*
Bracts, *27*, 88, *90*
Buddleja davidii, *86*
Butcher's broom, 67, *68*
Buttercup
 creeping, *115*
 Persian, 60, *64*
 water, 19, *21*
Butterfly bush, *86*

Calendula, 28, 29, *29*, 75, *80*
Calyx, *7*, 23–30, 65. *See also* Sepals
 relation to corolla, 31, *32*
Candolle, Augustin P. de, xxv
Canna, 36, *37*

Carnation, 31, 95, *96*
Carpels. *See* Pistil
Centaurea montana, 26
Chamaerops humilis, 17
Chrysanthemum grandiflorum, 9
Chrysanthemum morifolium, 8
Cirsium edule, 89
Citrus aurantiifolia, 18
Classification (botanical), xxii
Coen, Enrico, xxvi
Coltsfoot, *20*
Columbine, 48, *49*
Colutea, 74, *74*, 104n12
Compositae, 28–29
Composite flowers, 84–92
Contraction, 36, 75, 88. *See also* Expansion and contraction; Polarity
Coreopsis grandiflora, 32, 33
Cornflower, *26*
Corolla, *7*, 30–36, 48, 52, 95. *See also* Petals
Cotyledons (seed leaves), *7*, 10–16, 24, *25*
Crocus, 60
Crocus chrysanthus, 63

Daisy, *55*, *87*
Darwin, Charles, xxii, xxiv
Delicate empiricism. *See* Empiricism
Delphinium, 50, *51*, 108, *109*
Dianthus, 66
Dipsacus, 88, *90*
Dryopteris erythrosora, 69

Elm, 75, *77*
Emerson, Ralph Waldo, xxv
Empiricism (Goethe's "delicate"), xviii, xxiii, xxvi

Evolution, xxi, xxii, xxiv, xxv
Exact sensory imagination, xxviii, 108, 111
Expansion and contraction, xx, 31, 44, 48, 54, 60, 65, 74, 92, 100, 102. *See also* Contraction; Polarity
Eyes, 12, 22, 81–84

Faust, xv, xviii, xx
Fennel, xvii, *113*
Ferber, Johann Jacob, 97, 104n17
Fern, 69, *69*
Fevillea, 45
Flowering, 99–103
  effects of nourishment on, 23, 30, 88, 97
  transition to, 23, 28
Flower leaves, 26
Foeniculum vulgare, 113
Foliage leaves. *See* Stem leaves
Foliar theory, xxvi. *See also* Leaf (as basic plant organ)
Folia floralia, 26, 93
Fraxinus pennsylvanica, 78
Fruiting, 55, 56, 65–74, 84, 92, 99–103

Gaertner, Joseph, 83, 104n14
Gases (effects on metamorphosis), 19, 22, 74, 104n12
Genetic method, 105–115
Geoffroy Saint-Hilaire, Étienne, xxv
Gerbera jamesonii, 87
Gould, Stephen Jay, xxxn20
Grass, 88, *91*
Growth (vegetative), 55, 99–102

Haeckel, Ernst, xxii
Hedwig, Johannes, 22, 103n4

Helianthus annuus, 28
Helmholtz, Hermann von, xxiv
Herder, J. G., xxii
Honeysuckle, *57*
Humboldt, Alexander von, xxiv

Illustrations (Goethe's interest in), xxvii–xxix, 10, 92
Imagination, xviii, xxiii, 106. *See also* Exact sensory imagination
Intensification, xix–xx, 108
Intuitive perception, xviii, 111, 112. *See also* Sensory perception
Iris, 60, *61*, *72*
Iris pallida, 61
Italy, xv, xvii, xxii

Kant, Immanuel, 112
Kiggelaria, 45

Lactuca muralis, 113
Lathyrus odoratus, 53
"Law of environment," xx, xxi
"Law of inner nature," xx, xxi
Leaf (as basic plant organ), xvii, xviii–xix, xxv. *See also* Foliar theory
Leaf stalk (petiole), 16, *18*
Leaves. *See* Cotyledons; Flower leaves; Stem leaves
Leucanthemum x superbum, 55
Liber, 22
Light (effects on metamorphosis), 11, 19, 104n13
Lilium lancifolium, 59
Lily
    tiger, *59*
    water, *44*
Lime, *18*
Linden, 67, *67*

Linnaeus, Carolus, xvi–xvii, xxii, 45, 97–99, 103n7, 104n18
Lolium multiflorum, 91
Lonicera ciliosa, 57

Magnolia acuminate, 58
Maple, 75, *76*
Materialism (Goethe's rejection of), xxiii
Melianthus, 52, *52*
Metamorphosis, xv, xix
    accidental, 6, 10
    irregular (retrogressive), 6–7, *9*, 52, 60, 65, *66*
    regular (progressive), 6, *8*, 60
Metamorphosis of Plants (publication and reception of), xv, xxii, xxiii–xxix
Method, scientific. *See* Genetic method
Meyerowitz, Elliot, xxvi
Monarda didyma, 34
Monkshood, 50, *51*
Morphology, xvi, xxiv

Narcissus, 48, *48*
Nature Institute, xxxn21
Nectaries, 44–54, 56, *57*
Nectar, 45, 48, 56, *57*
Nerium, 48
Nigella
    damascena, 50, *50*, 70, *73*
    orientalis, 70, *73*
Nodes, 12, 16, 22, 23, *24*, 81, 82, 83, 88, *89*
Nourishment (effects on flowering), 23, 30, 88, 97
Nymphaea alba, 44

Owen, Richard, xxv

Padua (Italy), 17
Palm
    date, 16
    Mediterranean fan, xvii, *17*
Papaver
    atlanticum, *42, 43*
    orientale, *71*
    rhoeas, *40, 41*
Papilionaceous flowers, 52, *53*, 103n10
Parnassia, 45, *46*
Passiflora, *47*
Passion flower, 45, *47*
Pea, *53*, 103n10
Pentapetes, 45
Perception, intuitive. *See* Intuitive perception
Perception, sensory. *See* Sensory perception
Perennial plants (Linnaeus's focus on), 97–98
Petals, 5, 31, 35, *35*, 36, 48, *49*. *See also* Corolla
    relation to stamens, 36–45
Petasites
    frigidus var. *nivalis*, *20*
    frigidus var. *palmatus*, *20*
Petiole. *See* Leaf stalk
Pine, 14, *15*, 24, *25*, 85
Pinks, 65, *66*
Pinus
    contorta, *25*
    lambertiana, *85*
    nigra, *15*
Pistil, 7, 55–65, 92
Pitcher plant, *62*
Pith, 22, 98–99, 103n4

Poetry and science, xxiii–xxiv, xxvi, 111–112
Polarity, xix, xx, 108. *See also* Expansion and contraction
Pollen, 55–56
Polygala myrtifolia, 52, *53*
Poppy, 36, *40–43*, 70, *71*
Portmann, Adolf, xxvi
Potato, *82*
Primal plant, 104n23
Proliferous carnation, 95, *96*
Proliferous rose, 93, *94*
Proteus, xvii, xix, xxi, xxii, xxiv, 111

Ranunculus
    aquaticus (or *aquatilis*), 19, *21*
    asiaticus, 60, *64*
    repens, *115*
Reason (*Vernunft*) and understanding (*Verstand*), 111–112
Reproduction, 6, 55, 65, 82, 99–103
Ribs (of leaves), 16, 19
Richards, Robert J., xv, xxiv
Root, 82, 83, 104n13
Rosa damascena, *38, 39*
Rose, 36, *38, 39*, 93, *94*
Ruscus, 67, *68*

Salvia coccinea, *101*
Sap, 19, 22, 23, 28, 30, 31, 54, 74, 84, 100, 103n5
Sarracenia, 60, *62*
Scabiosa columbaria, *114*
Seedpods, 70, *71–74*
Seeds, 65, 69, 75, *76–80*, 83–84
Seed leaves. *See* Cotyledons
Sensory perception, xviii, xxiii, 106. *See also* Intuitive perception
Sepals, 30, 31. *See also* Calyx

Sicily, xvii
Sidalcea malviflora, 106, *107*
Solanum tuberosum, *82*
Spinoza, Baruch, xviii, 106
Spiral vessels, 54–55, 57
Stamens, *7*, 36–44, 54–56
    relation to petals, 36–45, *46*
    relation to style, 56–57, *58*, *59*, 60
Steiner, Rudolf, xxxn18
Stem, 22, 23, 102
Stem leaves (foliage), *7*, 14–22, 26, 31
    relation to calyx, 24, 26, *26*, 28, *28*, 29, *29*
    relation to corolla, 31, *34*, 35
Stigma, 60, *61–64*, 70
Style, 56–65
    relation to stamens, 56–57, *58*, *59*, 60
Subject/object dualism (Goethe's effort to overcome), xxiii, 111. *See also* Empiricism (Goethe's "delicate")
Suchantke, Andreas, xxxn21
Sunflower, 28, *28*

Teleology, xxi
Theory of Color, xvi
Thistle, 88, *89*
Thoreau, Henry David, xxv
Tilia platyphyllos, *67*
Trees. *See also names of various trees*
    Linnaeus's focus on, 97–99
Tulip (*Tulipa*), 35, *35*

Ulmus glabra, *77*
Understanding (*Verstand*) and reason (*Vernunft*), 111–112

Unity in diversity (Goethe's search for), xvi, xvii, xviii, xxii, 6, 54, 100, 102, 105–106
Urpflanze, xvii, xxvi. *See also* Archetype

Vallisneria, 45
Vegetative growth. *See* Growth (vegetative)
Vicia faba, *11*, 12, *12*, 13

Wall lettuce, *113*
Water (effects on metamorphosis), 19, *21*
Weimar, xvi–xvii, xxii
Wood, 98–99

Zannichellia, 60